Collins *gem*

Whiskies

Dominic Roskrow

First published in 2009 by Collins
an imprint of HarperCollins Publishers
77–85 Fulham Palace Road, London W6 8JB
www.collins.co.uk

Collins is a registered trademark of
HarperCollins Publishers Limited

12 11 10 09
6 5 4 3 2 1

A catalogue record for this book is available from
the British Library

Dominic Roskrow hereby asserts his moral right to be identified as the author
of this work.

This book was produced for HarperCollins by Thameside Media
www.thamesidemedia.com

ISBN: 978-0-00-729311-7

Printed and bound by Leo Paper Products Ltd,
China

Contents

Whisky Lore and Myths

It may be a popular drink, but whisky is also one that engenders more than a little reverence. Alongside this due respect for a fine drink, however, a few too many myths and fallacies have grown up about what's "right" and "wrong". So, before we go any further, let's debunk a few of the worst offenders.

Enjoying whisky
There will be few ground rules in this book, but the first and most important is that there is no right way of appreciating and enjoying whisky. If anyone tells you that you should add water to it, you shouldn't add water to it, you shouldn't add ice, you should drink it standing on one leg ... then politely tell them that this is a drink for individuals.

The blender's art

Most of this book will be concerned with single malt whisky because malt distilleries provide so many routes to explore and so many different flavours to discover. But it's worth bearing in mind that, although the demand for single malts is growing, blended whisky still accounts for about 19 out of every 20 glasses of whisky drunk. There are very good reasons for this.

Blended whisky tends to be tainted with a poor reputation because, frankly, a good proportion of it is rubbish. The average blend is made up of some malt whiskies mixed together and blended with a whisky made from another grain. Often this is mass-produced grain whisky with a fairly neutral taste, and it is added in large quantities, reducing the flavour of the malt. Often the grain is matured in tired old casks for the legal minimum time – three years. So, throw together high proportions of cheaply-made grain whisky and some second grade malt and you get the sort of cheap blends you find in small supermarkets, often with preposterously contrived Scottish-sounding names, on the lines of "Glen Kilt" and "Heather Bagpipe Macbeth Deluxe Scotch".

It doesn't have to be this way, and some of the most sophisticated and satisfying whiskies in the world are premium Scotch blended whiskies. They are some of the most expensive, too.

The skill of a master blender is one of the greatest artisanal skills, and it is a great mistake to dismiss blends as being in any way inferior to single malts. Many blends use only the finest malt whisky in

Chivas Regal 18-year-old blended whisky

relatively large proportions, blended with quality grain whisky produced in the finest oak casks.

Beyond Scotland

It simply isn't true that Scotland is the only great whisky producing nation. The sheer scale of production in Scotland, the large number of distilleries and the exacting quality controls ensure that, should a definitive list of the world's best whiskies be drawn up, it would certainly be dominated by Caledonia.

But on such a hypothetical list there would be many other distinctive whiskies from around the world. A few American bourbons, a smattering of

Distilleries such as Japan's Suntory are now acknowledged for making some of the world's best whiskies.

Irish whiskeys, and a number of Japanese and Canadian whiskies would not only be on the list, but would also be vying for the top spots.

Old, rare and expensive

Don't fall into the trap of thinking that older whisky is always better, and that whisky has to be expensive to be good. Although whisky becomes rarer and therefore more expensive after 20 years, it doesn't necessarily improve. The influence of the cask can begin to take over, and, just as with the flavour of oak in wine, it's not to everybody's taste.

Some whiskies are more robust than others, and cope well with the wood and taste fine well over 25 years of ageing, and even beyond 30 or 40 years. Such whiskies are, however, few and far between. Other malts are submerged by oak at a much younger age. So while whiskies aged for more than 25 years will require a serious investment financially, it's a moot point as to whether you'd want to drink them.

The flip side to this argument is, of course, that some whisky styles actually benefit from youthfulness, and are better drunk younger.

The bottom line is that you don't need to break the bank to enjoy outstanding whisky. We may all get the chance to taste a rare and old malt now and again, but, with only a modest amount of exploration and investigation, the whisky enthusiast is likely to rapidly build up a personal repertoire of value-for-money malts, undoubtedly including one or two reserved for the top shelf and that special occasion.

You say "whisky", I say "whiskey"

Finally, a word about the spelling: "whisky" or "whiskey". The word is an anglicised bastardisation of the Gaelic *uisge beatha* ("water of life") – words claimed from a language common to old Scotland and old Ireland. The Scots opt for the former spelling, the Irish favour the latter, with the "e".

In America "whiskey" is the more common spelling, though several producers, including Maker's Mark, choose to spell the word without the "e". Japan goes with the Scottish way, as does Canada.

Maker's Mark bourbon opts for the Scottish spelling of "whisky".

9

Types of Scotch

Although this book's primary focus is on malt whisky, it would be remiss of any book about the subject of whisky not to make reference to the full range of Scotch whiskies produced.

Malt whisky

Malt whisky is produced in batches, using only malted barley, yeast and water. The process of making malt whisky, from the malting of the barley through to maturation of the spirit, is covered in detail on pages 14–21. Malted barley isn't the only grain that can make whisky, and nor is the batch method the only way of distilling the wash. It is, however, the grain that ferments easiest with yeast because of the richness of the sugars it releases; it also produces the most flavoursome alcohol.

The Macallan
10-year-old single malt

Grain whisky

Whisky can be produced from a range of grains, including unmalted barley, corn, wheat and rye. These grains are always mixed with an amount of malted barley to ensure successful fermentation.

The fundamental difference between malt whisky and whiskies made with grains other than malt is that malt is made in batches, whereas other grain whisky production doesn't need to be. It can be made in a continuous, factory-like process in a continuous, or column, still.

Unlike the pleasantly curvy copper pot still, the column still is a workmanlike, no-nonsense piece of industrial equipment. Whisky is made in it by pouring fermented wash down a series of tall columns and against steam at extremely high temperature and pressure. The liquid is vaporised and then passes against a series of plates where it is condensed into liquid. Whereas malt comes off the still as a fruity, appley fresh spirit, grain spirit is considerably stronger and smells like soggy, slightly sweet

"Cameron Brig" single grain whisky

breakfast cereal in hot milk. The spirit it produces has far less taste than single malt, and is referred to in some countries as neutral grain spirit.

Grain whisky is bound by similar laws to single malt, and, in Scotland, that means it must be matured for a minimum of three years in oak casks. Almost all grain whisky is used to make blended whisky and very little is bottled as grain whisky in its own right. However, notable examples of single grain whisky do exist.

Blended whisky

The term blended whisky refers to a mix of malt whiskies combined with whisky made from another grain – usually corn or wheat, combined with a small amount of malted barley and possibly unmalted barley too. Quality blends can contain rare and old whisky, and can cost considerably more than most malts – so they are certainly not the poor relation of malt whisky.

Distillers tend not to disclose which whiskies go into their blends, however, which makes it difficult to anticipate

Bailie Nicol Jarvie
blended whisky

how a blend will taste until you've tried it. However, the big names of blended whisky have usually attained their position for very good reasons. Mainstream brands such as Johnnie Walker, Bell's, Famous Grouse, Dewar's, Whyte & Mackay and Teacher's use very good quality malt whisky from some of the most respected distilleries in Scotland. (See also pages 156–61.)

Blended malt whiskies (vatted malts)

If a blended whisky is the full orchestra of flavours, a vatted malt is the small ensemble of the whisky world – a harmonious blend of single malts from different distilleries. Vatted malts (now increasingly known as blended malts) may include anything from three different malt whiskies to a couple of dozen, but the mix contains only malt and no grain whisky.

Blended malts tend not to be restrained by traditional marketing baggage and several have adopted attention-grabbing names, such as Sheep Dip and Monkey Shoulder. (See also pages 161–3.)

Johnnie Walker Green Label blended malt

Making Malt Whisky

Single malt whisky is the type of whisky most closely associated with Scotland, and is the spirit produced by the vast majority of its 100 or so distilleries. Although the raw materials used in the production process are pretty much identical at every malt distillery, no two malt whiskies taste exactly the same. This is the mystery of malt: how three basic ingredients – water, malted barley and yeast – treated in so similar a way can yet produce such a panoply of flavours.

Fermentation

• The first stage of making malt whisky is to take the three basic raw ingredients of malted barley, yeast and water, and to make beer with them.

• Malted barley is barley that has been "tricked" into growing. To do this, the barley is soaked in warm water and left to germinate. After a while, the cell walls of the grain will break down, and a shoot will start to grow. The traditional way of doing this is to lay the malt out on large floors,

The traditional maltings at Balvenie Distillery

but the more common practice today is for distilleries to buy malt from commercial maltsters, who use automated systems.

• Germination releases starches and enzymes. The starches will become the sugars required to make alcohol, and the enzymes enable this process to occur.

• Germination has to be halted before it goes too far and destroys the starches. This is achieved by kilning. Malt is spread out on a metal floor above a furnace and dried by heat applied either directly from the furnace or indirectly by hot air that is passed through the malt.

• It is during this kilning process that smoke from burning peat may be employed to add flavour to the malt, eventually giving the final spirit phenolic, smoky and medicinal characteristics.

- After kilning, the malt is ground into a rough flour, or grist. This in turn is mixed with hot water in a mashing machine and then passed into a large holding vessel, known as a mash tun.
- The process at this stage is a little bit like making a pot of tea. Hot water is passed through the grain up to three times and this flushes out the enzymes and sugars required to make alcohol. The brown liquid – known as worts – passes out through the perforated bottom of the mash tun where it is collected and cooled.
- The worts are transferred into wooden or stainless-steel washbacks, where yeast is added and fermentation takes place. This process is the same as making beer, except that the fermentation isn't carried out in sterile conditions.
- The yeast effectively eats the sugars created by the starch and enzymes in the malt, producing alcohol and carbon dioxide. The process takes upwards of 50 hours and produces a sharp and sour beer due to the secondary fermentation provoked by bacteria. The result is a distiller's beer – known as wash. It has a strength of 5% to 9% abv (alcohol by volume), and this solution is now ready for distillation.

Distillation

Copper pot stills at Craigellachie Distillery

• Most, but not all, single malt whisky is distilled twice, in large copper kettles known as pot stills. Distillation is the process of heating the solution to boil off a variety of alcohols – all of which have different boiling points – and collecting the ones the distiller wants by separating them out and condensing them back from vapour to liquid.

• The first distillation takes place in the wash still. The still is heated, the vapours rise and pass up the copper still towards the tapered neck, then pass down a tube known as the lyne arm, where they are cooled in a condenser by cold water passing over the outside of the copper. The first distillation passes through the spirits safe into a receiving tank, known as the low wines, or feints, receiver.

• The liquid – known as low wines – will have an alcoholic strength of just over 20% abv. To prepare it for the second distillation, it is mixed with the stronger, rejected spirits (the foreshots) from the previous distillation run, giving it an alcoholic strength of about 28% abv.

• It is from the second distillation that the distiller will collect his final spirit, and so this time around he must carefully monitor the spirit and separate out the proportion of the run that will adversely affect the taste of the final alcohol.

• When the moment is right, the distiller will transfer the flowing spirit to a different holding tank. This "new make" spirit will be kept to make whisky; it forms the middle section, or heart, of the run. The longer the run continues, the weaker the spirit will become, and once more the distiller must decide when the "cut" is complete. At this point, he must transfer the flowing spirit into a third tank, where the rest of the run (the tails or feints) is rejected and passed back for the next distillation.

Casks of maturing whisky
at Glenfiddich Distillery

Maturation

New make spirit is clear and strong. It has estery flavours (young, yeasty, cerealy), strong fruit flavours (often green apples), liquorice and – in the case of whisky made using peated malt – a dense wood bonfire taste and smell. Water will be added to bring it down to a strength of around 63.5%. It is then put into oak casks.

Under Scottish law, spirit can be described as whisky only after it has been matured in oak casks for at least three years. There is, however, much scope for flexibility when it comes to the sort of casks used, and this will have a major impact on the flavour and quality of the final whisky when it eventually comes out of the cask.

By far the highest proportion of casks are made of American oak, but a significant proportion are made of European oak. European casks are about 10 times as expensive, but some distilleries make the outlay because they say it is essential to maintain the distinct nature of their whisky.

New oak casks would be too powerful for malt, and so in the vast majority of cases malt whisky is matured in casks that have been used for something else previously, most commonly bourbon or sherry, and that will greatly influence the flavour of the malt inside.

The mystery of casks

It is in the casks that the real magic of malt takes place. Casks come in a variety of sizes: barrels (180 litres), hogsheads (250 litres), puncheons (450 litres) and butts (500 litres). Smaller casks, such as quarter casks, are occasionally used also.

The cooperage at Balvenie

The reason for this is that the smaller the cask, the more liquid inside the container can come into contact with the wood, and therefore the more influence the wood can have on the whisky. Temperature is an intrinsic part of this, too, and Scotland's damp and cold climate ensures a stately and gentle maturation, lasting years.

Wood imparts colour and taste to the whisky. As the liquid warms, it expands into the wood. As it cools, it contracts away from the wood, drawing with it colour and flavour. The spirit will leave some fats and impurities in the wood, effectively removing some flavours from the whisky, and also the wood and the whisky will react with each other to create whole new flavours.

The more times the cask is used to make whisky, the more tired the wood will become and the less effective it will be at adding flavour. So when whisky is added to a cask for the first time (known as the first fill), the influence of sherry or bourbon will be at its strongest.

After three years in the cask, the spirit can legally be called whisky, and at some point after that it will complete its maturation and be removed from the cask for bottling or blending.

Making Malt Whisky

Nosing and Tasting

We talk about nosing whisky because much of the subtlety of a malt whisky is experienced through the nose, which can discern far more aromas than the tongue can taste flavours. It is important, therefore, to have a glass that will allow you to smell the contents when you are sampling whisky. The ideal glass will have a bulbous bottom and a tapered neck, narrowing to the rim so that the aromas are concentrated into a small area at the top.

Water or no water?
Don't let anyone tell you you shouldn't add water to malt whisky. In most cases, the malt in your glass has already been mixed with water, in fact,

to bring it down to bottling strength. If it has been bottled at cask strength instead, you'll need to add water in any case to bring it down to a tolerable level for drinking.

Adding water to malt whisky is the equivalent to a spring shower on a rose garden: it makes it far more fragrant. Blenders and whisky makers will nose spirit diluted to 20% abv. Adding water to take malt under 40% isn't for everyone, though.

Appraising

Before you nose or taste a whisky, you can get some valuable clues as to what to expect from its appearance. Shake a sample of whisky in a bottle. Standard strength (40–43% abv) whiskies will form big bubbles, which quickly disappear. A cask-strength whisky will form small bubbles, which remain longer. Once you've poured the whisky, swill it up the side of the glass. "Legs" – small columns of liquid – will form and run back down the glass. Slow, thick legs indicate an older whisky.

If the whisky has been matured in a bourbon cask, the colour will tend to be golden lemon and straw-coloured. Sherry casks impart an amber-mahogany colour. Deeper colours may indicate a

greater age, though relatively old whisky from a bourbon cask will be lighter than a much younger sherried whisky, and an older cask that has been used for whisky making before will impart less colour than a first fill cask.

Nosing

Swirl the whisky in the glass and then smell it from a distance, gradually bringing it closer to your nose. Nose the whisky with both nostrils and then one after the other – some find that one of the three ways of nosing suits them best.

What you're trying to do is identify familiar aromas in the whisky. Can you smell any citrus or berry fruits, is it spicy or smoky, can you smell sherry or vanilla? Don't worry if you can't identify anything in particular at first – it takes a while to get the hang of it. Whatever you identify is personal to you, so, if you think you smell roast beef and Yorkshire pudding topped with English mustard, write it down and stick with it. As long as it's a smell you would recognise again, it's a suitable descriptor for you.

Tasting

When it comes to tasting, don't sip the whisky but put a sizeable amount of it in the mouth and hold it there before swallowing. You can spit it out, but, unless you're going to be tasting a lot of whisky, there's no need to do this. From the taste, you want to try and assess how it feels in the mouth – rich and mouth-filling or thin and winey? Is it zesty or fizzy in any way? Does it taste like a whisky, and does it taste good? Does it linger in the mouth when you've swallowed it? The first three questions constitute the "palate", while the fourth refers to the "finish".

Take notes

Write down the name of the malt, its strength and whether you like it or not. If it reminds you of anything, write that down too. These notes can take whatever form you like, but it's best to divide them into: colour, nose, palate and finish.

Buying Whisky

Entering a specialist whisky shop can be a little intimidating, especially if it's fronted by a conceited retailer who senses fear the minute the shop door opens. Fortunately, many retailers these days are choosing to discard the funereal outfits, dispense with the semi-darkness, turn up the lights and share their passion for whisky.

Bottle labels

Labels on whisky bottles can confuse and confound. The only essential information, however, is the name of the distillery, the age of the whisky, its strength and its price. All else is

This label tells us that the whisky – named Authenticus – is a single malt from Benriach Distillery, is peated (giving it an earthy quality), has been aged in oak casks for at least 21 years and, at 46% abv, is a little stronger than most.

inessential – although information on the region
that the whisky is from may offer a guide to the
style, as will tasting notes, if given.

Age statements

The age on the bottle of a whisky is a rough guide
to its quality, and it will strongly influence what
you are going to pay for a bottle. It is only a guide,
though, and it doesn't always follow that older
means better. The age refers to the youngest
whisky in the mix. It is not an average, and nor
does it mean that all the whisky is that age –
there may be much older whisky in the malt.

If a producer accidentally adds one drop of a
younger whisky the whole batch has to be sold at
the age of that one drop. It does happen. A few
years ago the owners of Ardbeg accidentally
added some very young whisky from another of
their distilleries to some very old and rare Ardbeg,
not only destroying the age but preventing it
from being called a single malt any more. But the
mix tasted so good that they bottled it under the
name Serendipity without an age on the label.

This brings us to the next point. If a whisky
doesn't have an age statement, it doesn't follow

that it is cheap and nasty, or indeed, inferior at all. Johnnie Walker Blue Label, for example, is one of the most expensive blends on the market, but has no age statement. This is probably because a small amount of youthful, zesty whisky has been added to the mix to stir up the very special and rare whiskies it otherwise contains. Under the rules, even a drop of 10-year-old whisky would make the blend a 10-year-old, and would do a disservice to the quality aged whisky that dominates the drink.

Non chill-filtering

Making the perfect malt whisky is a balancing act. As far as possible, the aim is to remove bad tasting flavour compounds but leave in the good ones, and this is the crux of the non chill-filtering debate. When malt whisky is made, a large number of oils and congeners (organic chemicals such as esters, acids and aldehydes) pass into the final liquid. When the liquid is cooled, they naturally solidify and turn the whisky cloudy.

A non chill-filtered Ardmore single malt

This has traditionally been considered unattractive to producers, who want their whisky to be bright and clear. To ensure this, the liquid is chilled until the fats and impurities have solidified, and then the liquid is passed through a barrier filter to remove the solids from the whisky.

In recent years, however, there has been a lobby of enthusiasts who have argued that the congeners contribute to the full flavour of the malt and that, by removing them, the whisky is left clear but also a little bit blander.

In response, some producers have stopped chill-filtering, preferring to sacrifice appearance for taste. Indeed, they have made a virtue of non chill-filtering.

Cask strength

After distillation, new make spirit will tend to have an alcoholic strength of somewhere in the high 60s% abv (alcohol by volume). Most distilleries put spirit into cask at a strength slightly lower than that, so water is added at the outset. In Scotland, that

Aberlour's cask strength a'bunadh whisky

strength will gradually decline while maturing as some of the spirit evaporates (known as the "angels' share"), but, even after maturation, the cask whisky may still have a strength in the mid or even high 50s% abv.

Most malt whiskies are bottled with a strength of 40%, 43% and occasionally up to 46% abv. To reach these levels, the cask whisky is watered down for bottling. These days, however, it is quite common to find versions of top name whiskies bottled at the strength they came out of the cask. Caution is advised when drinking such whiskies, and even on the nose the acerbic burn of alcohol is pronounced. It is preferable to add small amounts of chilled water to cask strength whiskies to make them drinkable.

The benefit to the drinker is that the added water can release the aromas and flavours of the whisky, but it can still be consumed as a normal full-strength whisky. It also means that, effectively, you are getting up to half an extra bottle of whisky for your money.

And, remember, if anyone ever tells you that you shouldn't add water to malt whisky, point out that, on a standard bottle, someone already has!

Special wood finishes

Although it is not permitted to add anything to the basic mix of barley, water and yeast and call it Scotch whisky (with the exception of a small amount of caramel for colouring consistency purposes), the rules do allow the whisky to be matured in casks that have contained something else. Indeed, a pre-used cask is essential to the development of Scottish malt.

A common practice among many distillers is to take a whisky out of the oak barrel that it has spent some years in, and finish its maturity in a totally different cask. The time the whisky is in the second cask can be anything from a few weeks to two years or longer.

There have been experiments with many types of cask, including rum, Madeira, port, burgundy, Champagne and claret, and there have been a number of "pink" whiskies as a result of the process.

Balvenie Doublewood is aged in American ex-bourbon barrels, then finishes its maturation in European ex-sherry casks.

The cask also imparts some flavour, of course – usually in the form of fruitiness. This process can enhance malt and produce some excellent results, but don't entirely accept the marketing spiel on the label. The process isn't without its critics, who argue that cask finishing is often used to disguise poor quality malt.

Independent bottlers

Official distillery bottlings are to whisky what official CD releases of a rock band are to music. And it follows that the releases from independent bottlers are the industry's "bootlegs" – unofficial releases that fans must have. Like bootlegs, their quality varies from utter gems to sub-standard versions of all-time classics.

Every official bottle from a distillery is a careful marriage of casks, with the marriage designed to ensure that, no matter when or where the whisky is purchased, to all intents and purposes, it tastes the same. But most single malt whisky is produced to be used in blended whisky, and it is common practice for different distilleries to swap malts so that each company has the widest selection from which to create their blends.

Inevitably – and particularly when whisky is in one of its ebb periods – some casks of malt become available on the open market, and it is these that the independent bottlers seize upon. Unlike the rounded and balanced mix of casks that the distillery's owner will put out officially, an independent bottler might get hold of just one, two or three casks at a time. The malt might have been aged for an unusual number of years, and, as each and every cask differs from the next, the whisky that ends up inside the bottle can differ markedly from what might be expected.

This is part of the appeal, of course. If a cask is going to yield only 250 to 300 bottles, a single cask bottling will be extremely limited, and that gives it rarity value. If you buy an independent bottling from your favourite distillery, there is always the potential to find your perfect dram.

This independently released whisky called Ledaig is a bottling of Tobermory whisky and has an unusual age statement of 9 years.

Scotland's Distilleries

Scotland's malt whisky tends to be classified into the following broad regional and stylistic divisions:

Speyside: Home to about a third of Scotland's distilleries. The whiskies are complex, sophisticated and prized for their blending potential; the sweet end of the whisky spectrum.

Highlands: Full-bodied, rich and robust whiskies, with a complex array of flavours; often smoky and with an earthy, peaty quality.

Islands: A variety of styles, from clean and citrussy (Jura, Scapa) to bold and peated (Talisker, Highland Park).

Islay: Normally the most intense of whiskies: heavily peated and briny, with tar-like qualities. But Bunnahabhain and Bruichladdich demonstrate that the island has other styles too.

Lowlands: Light-bodied and usually light-hued, with grain, grassy and delicate floral notes.

Scotland's whisky regions

ISLANDS

Orkney

John o' Groats

Durness

Stornoway

Outer Hebrides

Ullapool

SPEYSIDE

INVERNESS

Dufftown

Kyle of Lochalsh

Grantown-on-Spey

ABERDEEN

Skye

HIGHLANDS

Rum

Mallaig

Fort William

ISLANDS

Tobermory

Mull

Oban

PERTH

DUNDEE

St Andrews

Jura

STIRLING

Inveraray

Kintyre

EDINBURGH

Berwick-upon-Tweed

GLASGOW

ISLAY

Arran

Lanark

Jedburgh

Campbeltown

LOWLANDS

Dumfries

Stranraer

Scotland's malt distilleries by region

The distilleries are listed in alphabetical order on pages 38–151, while below we list them within the principal Scottish whisky regions.

Speyside

Aberlour
Allt-a-Bhainne
Auchroisk
Aultmore
Balmenach
Balvenie
Benriach
Benrinnes
Benromach
Braeval
Cardhu
Cragganmore
Craigellachie
Dailuaine
Dufftown
Glenallachie
Glenburgie
Glendronach
Glendullan
Glen Elgin
Glenfarclas
Glenfiddich

Glenglassaugh
Glen Grant
Glenlivet
Glenlossie
Glen Moray
Glenrothes
Glen Spey
Glentauchers
Huntly
Inchgower
Kininvie
Knockando
Linkwood
Longmorn
Macallan
Macduff
Mannochmore
Miltonduff
Mortlach
Roseisle
Royal Brackla
Speyburn
Speyside

Strathisla
Strathmill
Tamdhu
Tamnavulin
Tomintoul
Tormore

Glenfarclas Speyside
single malt whisky

Highlands
Aberfeldy
Ardmore
Balblair
Ben Nevis
Blair Athol
Clynelish
Dalmore
Dalwhinnie
Deanston
Edradour
Fettercairn
Glencadam
Glen Garioch
Glengoyne
Glengyle
Glenmorangie
Glen Ord
Glen Scotia
Glenturret
Knockdhu
Loch Lomond
Oban
Old Pulteney
Royal Lochnagar
Springbank
Teaninch
Tomatin
Tullibardine

Islands
Arran
Highland Park
Jura
Scapa
Talisker
Tobermory

Islay
Ardbeg
Bowmore
Bruichladdich
Bunnahabhain

Caol Ila
Kilchoman
Lagavulin
Laphroaig
Port Charlotte

Lowlands
Auchentoshan
Bladnoch
Daftmill
Girvan
Glenkinchie

Talisker malt from the Isle of Skye

Glenkinchie Lowland malt

37

Aberfeldy <small>HIGHLANDS</small>

Aberfeldy, Perthshire
www.aberfeldy.com
www.dewarsworldof
  whisky.com

Core range
Aberfeldy 12-year-old
Aberfeldy 21-year-old

Signature malt
Aberfeldy 12-year-old:
rich, oily and fresh, with a
zesty tangerine note

The Aberfeldy malt is very much underrated. It's
a rich, honeyed and oily whisky, with full flavours
– particularly in the 21-year-old. The distillery and
its associated Dewar's World of Whisky exhibition
are set in wonderful surroundings and easy to
reach. A tour of the pretty distillery is worthwhile
for the more serious whisky enthusiast, while
Dewar's World of Whisky focuses on the Dewar's
blend, which has the Aberfeldy malt at its heart.

Aberlour SPEYSIDE

Aberlour, Banffshire
www.aberlour.co.uk

Core range
Aberlour 10-year-old
Aberlour 12- and 16-year-old
  Double Cask Matured
a'bunadh

Signature malt
Aberlour 10-year-old:
surprisingly weighty and full,
with honey and malt in balance
and just a hint of the distillery's
trademark mint notes

Nestled at the end of Aberlour's busy main
street, Aberlour distillery remains largely as it was
when completed by architect Charles Doig at the
end of the 19th century. You can tour Aberlour to
sample the wort and the wash as well as the
finished whisky – which is a rich, satisfying dram.
Each batch of a'bunadh is slightly different.

Allt-a-Bhainne SPEYSIDE

Glenrinnes, Dufftown, Banffshire
www.pernodricard.com
www.chivas.com

Seagrams built Allt-a-Bhainne along
with Braeval Distillery in the 1970s.
The construction was to help meet
the growing demand for blended
whisky, and Allt-a-Bhainne has only
ever been a provider of malt for
blends. Consequently, its fate has
been tied to that of blended whiskies per se.
It was mothballed in 2002 but reopened three
years later when current owner Pernod Ricard
needed more malt to expand output of its newly
acquired Chivas Regal brand.

Allt-a-Bhainne is a large distillery, producing
an estimated four million litres of spirit each year,
and is designed to operate with the minimum
number of people. There are no official malt
bottlings, but some malt is occasionally bottled
independently.

Display of whiskies at Ardbeg (see next pages)

Ardbeg <small>ISLAY</small>

Port Ellen, Isle of Islay
www.ardbeg.com

Core range
Ardbeg 10-year-old
Ardbeg 17-year-old
Kildalton
Provenance 1974
Uigeadail
Airigh Nam Beist
Lord of the Isles

Signature malt
Ardbeg 10-year-old: a meal in a glass;
cocoa, oily fish, swirling peat and chewy sweetness
married together perfectly; truly exceptional

South Islay's whiskies are famous for their
distinctive peaty, smoky style, and Ardbeg is one
of the "big three" distilleries that sit next to each
other in what might be called whisky nirvana.
The sea laps across the rocky shoreline right up
to the distillery walls, and arguably there are few
experiences finer than drinking malt here,

straight from the cask on a blustery and sunny Islay day. If peaty, tangy, tarry and oily whisky is your thing, you will find Ardbeg to be sublime.

In addition to the core range, cask strength bottlings and a few oddballs are also on offer. Very Young, Still Young, Almost There and Renaissance track the maturation of the signature 10-year-old from six years up to full maturity. There are also some fantastic special bottlings originally casked before 1980. Be warned though: Ardbeg can easily become a very expensive lifelong pursuit.

The word "quaint" might have been invented for Ardbeg, and a ramshackle tour here takes you past hand-painted signs and a medley of traditional distilling equipment. It ends in what used to be the kiln for malting barley and is now one of the finest cafés in Scotland.

Scotland's Distilleries

Ardmore  HIGHLANDS

Kennethmont,
Aberdeenshire

Signature malt
Ardmore Traditional Cask: lightly
peated, non-chill filtered and
bottled at 46% abv to maximise
the natural flavours

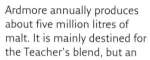

Ardmore annually produces
about five million litres of
malt. It is mainly destined for
the Teacher's blend, but an
increasing amount is released as a single malt.
As at sister distillery Laphroaig, some of
Ardmore's whisky is matured in quarter-sized
casks to develop the malt flavour further.

As for the whisky, Ardmore Traditional Cask is
a peated malt that enjoys the highest quality
maturation – first in ex-bourbon barrels and then
in traditional quarter casks.

Arran <small>ISLANDS</small>

Lochranza, Isle of Arran
www.arranwhisky.com

Core range
Arran 10-year-old
The Arran Malt
Arran 100 proof
Robert Burns Single Malt
Cask bottlings

Signature malt
Arran 10-year-old: creamy, rich mix of citrus, toffee and
butterscotch; very chewy and quite delightful

Arran has something of a scattergun approach to
releasing malts, and in the early days released
some ropey young whiskies. But by the time the
distillery was bottling a 10-year-old it had turned
into a glorious swan of a malt. Its rich creaminess
is credited to its location at Lochranza, where it
sits in a suntrap in the Gulf Stream. Recent cask
strength and non chill-filtered versions of the malt
are particularly impressive. Also keep an eye out
for one of the occasional special cask finishes.

Auchentoshan Lowlands

Clydebank, Glasgow
www.auchentoshan.com

Core range
Auchentoshan Select,
Auchentoshan 12-, 18- and 21-year-
olds
Auchentoshan Three Wood

Signature malt
Auchentoshan 12-year-old: light and
smooth, with a hint of citrus; very
accessible to the beginner

Auchentoshan is special because it is a triple
distilled single malt. It bears the trademark
characteristics of a Lowland malt in that it is light
and easy to drink. But some of the more recent
bottlings show a surprising and impressive
diversity. It's worth comparing the clean, subtle
and citrussy 18-year-old alongside the Three
Wood, for example – the latter drenched in sherry
flavours. The Limited Editions are fantastic
examples of the very best Auchentoshan casks.

Auchroisk SPEYSIDE

Ulben, Banffshire
www.malts.com

Signature malt
Auchroisk 10-year-old

Special occasions
Auchroisk 28-year-old Rare Malt

Producing 3.1 million litres each
year, Auchroisk is a relatively
modern distillery, built in 1974
to supply what was then the
International Distiller and Vintner group
(IDV) with malt for blending.

   The choice of location was taken after careful
consideration and so, perhaps unsurprisingly,
the distillery's malt proved to be of high quality.
Some of it has been bottled as a single malt
since the mid-1980s. Early bottlings were
known as The Singleton, but the distillery's
name is now used.

Aultmore SPEYSIDE

Keith, Banffshire

Signature malt
Aultmore 12-year-old

Only a tiny fraction of
the whisky produced at
Aultmore is bottled as a
single malt, with the
vast majority being used
for blending purposes.
Unsurprisingly, then,
this is a clean, consistent, no-nonsense malt.
The distillery started producing in 1897 and has
been in demand pretty much ever since.

There may not be much Aultmore around
as single malt, but if you are interested in
exploring this whisky further, it would be worth
seeking out some independent bottlings, such
as those produced by Elgin-based bottlers
Gordon & MacPhail.

Balblair HIGHLANDS

Edderton, Ross-Shire
www.inverhouse.com

Core range
Vintage 1997
Vintage 1989
Vintage 1979

Signature malt
Vintage 1997: this is classic
Balblair – full-bodied, clean, rich and fruity. In the older vintages, flavours include apricot, orange, green apple, cloves, vanilla and honey

Balblair has worked to reposition the distillery's malts to focus on vintage bottlings rather than the standard age expressions. That, along with impressive premium packaging, suggests Balblair is a malt that's going places. They say that the air around the distillery is the purest in Scotland, and Balblair associates itself with this purity. But Balblair isn't a lightweight, and has some distinctive earthy and spicy flavours that make it an attractive and satisfying malt.

Balmenach SPEYSIDE

Cromdale, Grantown-on-Spey

Balmenach produces whisky exclusively for blending, so there are no single malt bottlings released from the distillery – though the owner Inver House qualifies the above statement with the words "as yet", so the situation may change in the future. It's a sizeable operation, nevertheless, with capacity for the production of two million litres of spirit each year, some of which finds its way into independent bottlings.

BALVENIE
12729 LITRES
SPIRIT STILL Nº 2

Pot stills at Balvenie (see next pages)

Balvenie SPEYSIDE

Dufftown, Banffshire
www.thebalvenie.com

Core range
DoubleWood 12-year-old
Signature 12-year-old
Single Barrel 15-year-old
PortWood 21-year-old
30-year-old
Cask 191 50-year-olds
Vintages and Limited Editons

Signature malt
DoubleWood 12-year-old: chewy rich
fruits and the most exquisite Speyside
honey

Special occasions
The 21-year-old is quite possible the best example there
is of a whisky finished in port pipes, but the 30-year-old
is a world-class big hitter, dressed in ermine

Balvenie is the sister distillery to Glenfiddich, the
world's biggest malt distillery. What it lacks in
scale and quantity, Balvenie more than makes

up for in quality, with its fruity, honeyed whiskies.

Balvenie is situated on the same site as Glenfiddich and owner William Grant's third distillery, Kininvie. Balvenie's output is very much as a whisky lover's whisky, its distinctive toffeeness earning it a reputation as one of Speyside's top whiskies. It is the perfect foil to the glitzy Glenfiddich: a traditional craft distillery that does things the old-fashioned way. New expressions include the Signature 12-year-old, while Vintage Editions, such as the 1972 Vintage Cask are worth seeking out.

A visit to Balvenie includes a look at one of the few remaining floor maltings in existence and ends with a tasting session in the old distillery manager's cottage.

Ben Nevis  HIGHLANDS

Lochy Bridge, Fort William
www.bennevisdistillery.com

Signature malt
Ben Nevis 10-year-old

Special occasions
Ben Nevis 13-year-old
Port finish

The stunning location and a
sympathetic refurbishment by its
Japanese owners draw visitors to
Ben Nevis. The distillery has had a chequered
past, but is now approaching its 20th anniversary
since the acquisition by Japan's second largest
distiller, Nikka, and a steady flow of quality 10-
year-old and the occasional special bottling have
established it as a solid malt producer. Ben Nevis
is one of the very few distilleries that bottles a
single malt and a blend under the same name, so
check what you're buying. The distillery also sells
its own Glencoe blend, which is eight years old.

BenRiach SPEYSIDE

Longmorn, near Elgin,
Morayshire
www.benriachdistillery.
co.uk

Core range
Heart of Speyside
Curiositas
Authenticus
Hereditus Fumosus
Authenticus Fumosus
Arumaticus Fumosus
Benriach 15-year-old

Signature malt
Heart of Speyside 12-year-old: supreme Speyside
character – all rich fruit and honey, held in place by a
balanced oak and malt lining – exquisite

BenRiach was taken on by a consortium headed
by Billy Walker in 2004, since when it has gone
into whisky overdrive. Fortunately, most of the
whiskies are excellent. For a treat, try the
Authenticus. Aged for 21 years, this is a growling
beast of a peated whisky.

Benrinnes SPEYSIDE

Aberlour, Banffshire
www.malts.com

Signature malt
Benrinnes 15-year-old

This is another of Diageo's production distillery, producing malt for a range of blends, and with little whisky released as single malt. It produces a healthy 2.6 million litres of spirit each year.

There are technical points of interest at Benrinnes. It is one of just a handful of distilleries using the "worm tub" method to condense the spirit. The method is so named because of the worm-like horizontal pipes which lie in a tank of cool, flowing water.

The other unusual feature is that its six stills are arranged in two groups of three, with one wash still feeding two spirits stills.

Benromach

Forres, Moray
www.benromach.com

Core range
Benromach Traditional, Organic
 and Peat Smoke
Tokaji
Sassicaia
Portwood
Benromach 25-year-old

Signature malt
Benromach Traditional: a delicate mix of
malt and fruit with a spicy afterglow and
a trace of smoke

Special occasions
Benromach 21-year-old: a heady mix of sherry, dark
fruits, oak and spice – exquisite

Owner Gordon & MacPhail knows the mysteries
of wood management as well as anyone else in
the industry, so it should come as no surprise to
learn that Benromach is starting to produce
some excellent wood-finished malts.

Scotland's Distilleries

Bladnoch LOWLANDS

Wigtown, Wigtownshire
www.bladnoch.co.uk

Core range
Bladnoch 10-, 12-, 13- and 15-year-olds

Signature malt
Bladnoch 10-year-old: almost defines the Lowland style, with floral and light citrus notes and a complex mix of other influences

Special occasions
Bladnoch 15-year-old: the cask strength is chewier, more intense and deeper than the normal bottlings

Bladnoch is a very small, isolated and pretty distillery, and produces a style of whisky that for a long time was deeply neglected – the light, floral, Lowland type. Despite its size, Bladnoch offers tours and tastings. About 25,000 visitors make their way to the distillery each year, and few of them leave disappointed.

Blair Athol

Pitlochry, Perthshire
www.discovering-distilleries.com

Signature malt
Blair Athol 12-year-old:
unfussy and richly fruity

Special occasions
A very rare 27-year-old is worth
investigation – if you can find a bottle

Blair Athol lies close to the A9, not
far from Edradour Distillery, in the
region of the Highlands to the
south of Speyside. It is a sizeable distillery,
capable of producing about two million litres a
year. But very little of this is bottled as a single
malt, and it has just one core expression, the
Blair Athol 12-year-old. The vast majority of its
malt output goes into the heart of Bell's
blended whisky.

The distillery is one of the oldest working
distilleries in Scotland, having been established
in 1798, a century before many others close to it.

Bowmore ISLAY

Bowmore, Isle of Islay
www.bowmore.com

Core range
Bowmore 12-, 15-, 18-
and 25-year-olds

Signature malt
Bowmore 12-year-old: the classic Bowmore, with a
lovely balance of oak, malt, sea notes and mid-range
peat smoke

Special occasions
Bowmore 18-year-old: this is a delight. The balance of
floating smoke, fruit and oak wrap around the
distinctive and chunky malt perfectly

Bowmore is in the middle of Islay, and in terms
of peating levels, its whiskies are in the central
ground too, with the distinctive Islay taste but
without the phenol and fish peaks of some of the
other Islay whiskies. In the past, the distillery
suffered from producing too many different
styles of whisky, so the owners are now focusing

on a smaller core range. It's a beautiful whisky, with a distinctive "parma violet" note on some expressions. The 18-year-old has replaced the much loved 17-year-old, but its mid peat and oak hit hold up well.

Bowmore is a wonderful place to visit too. It improved its facilities a couple of years ago and now has a visitor centre that boasts stunning views across Loch Indaal to Bruichladdich Distillery. When the breeze is up and the sun flits across the busy waves that lap up to the distillery, take a glass of Bowmore and drink it whilst sitting on the sea wall – you'll never feel more alive!

Bowmore also has its own floor maltings and huge peat-burning fires, so you can see and smell the work in progress.

Bruichladdich

Bruichladdich, Isle of Islay
www.bruichladdich.com

Core range
Bruichladdich 10-, 12- and
15-year-olds
Bruichladdich XVII

Signature malt
Bruichladdich 10-year-old: clean,
unpeaty, sweet and fruity; very
moreish

Special occasions
Bruichladdich XVII: very fresh – almost zesty – and on its
best behaviour, but with enough bite for interest

Bruichladdich reopened its doors in 2000 when
it was bought by independent bottler Murray
McDavid. Traditionally it has tended to play the
role of maverick, portraying itself as a people's
champion – a David against the industry goliaths.
Even the style of whisky – clean, fresh and very
fruity – breaks the ranks on Islay.

With the highly respected Jim McEwan in charge of production and a swashbuckling approach to whisky making that has seen this Islay distillery create all sorts of eyebrow-raising concoctions, there's rarely been a dull moment at the distillery. A knowledge of all the Bruichladdich bottlings over the last few years would make a great subject for an appearance on *Mastermind*.

Tours of the distillery can take place throughout the year, but you are advised to ring in advance to book. There is a small charge, but the price includes a dram of whisky. The distillery also has a shop. The whiskies have great packaging.

Bunnahabhain

Port Askaig, Isle of Islay
www.bunnahabhain.com

Core range
Bunnahabhain 12-, 18- and 25-year-olds

Signature malt
Bunnahabhain 12-year-old

Special occasions
Bunnahabhain 25-year-old: a weighty whisky, with rich plum and sherry notes

Pronounced "Boo-na-ha-venn", Bunnahabhain is now in the ownership of Burn Stewart, who has been promoting the whisky as the "gentle taste of Islay". In other words, it is an unpeated whisky from the "peat capital" island of Islay.

Caol Ila ISLAY

Port Askaig, Islay
www.malts.com

Core range
Caol Ila 12-, 18- and 25-year-olds
Caol Ila Cask strength

Signature malt
Caol Ila 12-year-old: oily, with a seaside
barbecue combination of smoky bacon
and grilled sardines

Caol Ila is the biggest whisky
producer on Islay. It's not the best
known, however, as most of its malt
goes into blends, particularly Johnnie Walker's.

In 2001 – and partly because of stock
problems at owner Diageo's other peated Islay
whisky Lagavulin – Caol Ila started to be sold as
a single malt in three expressions. It has since
become the island's fastest growing malt –
which is no wonder, as this is a truly special
whisky, and the 18-year-old, in particular, is up
there with the very best.

Cardhu <small>SPEYSIDE</small>

Aberlour, Banffshire
www.malts.com

Signature malt
Cardhu 12-year-old: sweet,
very malty, very clean
and very drinkable

Cardhu is the symbolic home
of Johnnie Walker, and its
malt is a main component in
the range of Walker blends.
But, for all its high-profile associations, Cardhu
exists in a whisky limbo-land. It enjoys a huge
market in Southern Europe, particularly in Spain,
and is much in demand for blending. In certain
circles, it is even regarded as malt at its very finest.
Yet it receives none of the acclaim in its home-
land that's usually reserved for great Speysiders;
it attracts a relatively small number of visitors
too. More's the pity, as it is a charming distillery.

Clynelish HIGHLANDS

Brora, Sutherland
www.malts.com

Signature malt
Clynelish 14-year-old

Clynelish is rather enigmatic. It
has the characteristics of both a
seaside malt and a Highland one.
Furthermore, it is situated in the
town of Brora, next to another
distillery that was originally called
Clynelish but changed its name to
Brora. For a short time, the two
distilleries operated side by side as Clynelish
1 and 2, before the older one closed.

Clynelish now produces a rich and smoky
malt, and is highly recommended. But it's
nowhere near as peaty as the original Clynelish
style and the other whiskies produced at Brora.

Cragganmore SPEYSIDE

Ballindalloch, Banffshire
www.discovering-distilleries.com

Signature malt
Cragganmore 12-year-old:
complex and rich Speyside
fruit with a much less typically
Speyside tangy undertow

Special occasions
Cragganmore 17-year-old: bottled at
cask strength and limited to a few
thousand bottles

Collins Gem Whiskies

Cragganmore, one of the smallest
distilleries owned by drinks corporation Diageo,
is a sophisticated sweet-and-sour fruit mix of
a whisky. It is one of Diageo's six "classic malts",
representing the Speyside region in that
collection, though it is not entirely typical
of Speyside in its character.

Craigellachie SPEYSIDE

Craigellachie, Aberlour

Signature malt
Craigellachie 14-year-old: a nice mix of malt spice and fruit. Unchallenging but perfectly palatable

The village of Craigellachie is situated in the very heart of Speyside, alongside the rivers Spey and Fiddich. For many years the main hotel in the village has been seen as a base camp for all good whisky expeditions. The Craigellachie distillery itself is not very exciting, though – its featureless glass front and garish red lettering are suggestive of a factory rather than a malt producer. Very little Craigellachie is bottled as single malt, most being used by Dewar & Sons.

Dailuaine SPEYSIDE

Aberlour, Banffshire
www.malts.com

Signature malt
Dailuaine 16-year-old

Founded in 1852, Dailuaine has
been in almost continuous
production for 155 years, except
for three years between 1917
and 1920, when it was closed
due to fire damage. With the
potential to produce more than
three million litres of spirit a
year, Dailuaine is one of
Scotland's biggest malt contributors, yet one of
its least known. That is because only a small
percentage of the spirit made here makes it into
single malt bottlings; most is used for Johnnie
Walker blends, such as the classic Black Label.

Dalmore HIGHLANDS

Alness, Ross-shire
www.thedalmore.com

Core range
The Dalmore 12- and 15-year-olds
Gran Reserva
1263 King Alexander III

Signature malt
The Dalmore 12-year-old: muscular,
with orange notes and a solid oak
and malt platform

Special occasions
The Dalmore 1263 King Alexander III: stunning and
complex mix of bourbon notes, sherry and rich fruit

There are some who believe that Dalmore is up
there with Dalwhinnie and Clynelish as one of
the very best Highland distilleries. The whisky has
hit the headlines in recent years because one or
two of its very oldest whiskies – more than 60
years old – have fetched tens of thousands of
pounds at auction.

Dalwhinnie HIGHLANDS

Dalwhinnie, Inverness-Shire
www.malts.com

Signature malt
Dalwhinnie 15-year-old: whisky's
answer to a Harlan Coben crime
novel, twisting its way into and
out of taste cul-de-sacs at
breathtaking pace, before reaching
an unexpected but totally satisfying
climax. Earthy, smoky, swampy,
overwhelming – a thoroughly
recommended whisky

Dalwhinnie is in the heart of the Highlands, in
what is officially the highest spot for a distillery
in Scotland at an altitude of a little over 1,000 feet.
It is surrounded by the Grampians and the
Cairngorms, and overlooks tributaries to the
Spey. It is one of Diageo's original six "classic
malts", and it is just that – one of the truly great
Highland malts.

Deanston HIGHLANDS

Deanston, nr Doune, Perthshire

Signature malt
Deanston 12-year-old: reliable rather than flashy, with clean honey and malt notes

Special occasions
Deanston 30-year-old Single Malt Limited Edition: a veritable old gent – the years in cask have given it great depth, with a tangy, spicy edge

Deanston Distillery is capable of producing about three million litres of whisky a year, with much of the output going into the respected Scottish Leader blended whisky. Less than a fifth of the spirit ends up as a single malt and – not to put too fine a point on it – there hasn't been much to write home about it in the past. If ever there has been a "journeyman malt", this has been it. However, the signs are that the 12-year-old is on the up, and some experts have observed a marked improvement.

Dufftown SPEYSIDE

Dufftown, Keith, Banffshire
www.malts.com

Signature malt
Flora & Fauna 15-year-old

Special occasions
Singleton of Dufftown

Diageo's biggest producer at Dufftown makes about four million litres of whisky each year, the majority destined for Bell's. The distillery has been expanded several times in the last 30 years, and it has one of the longest fermentation periods of any whisky – up to 120 hours. Only a very small amount of whisky from Dufftown Distillery is ever released as single malt.

Edradour HIGHLANDS

Pitlochry, Perthshire
www.edradour.co.uk

Signature malt
Edradour 10-year-old

Special occasions
Edradour 30-year-old

Edradour is independently
owned by the bottler Signatory,
which is headed up by Andrew
Symington. The distillery has
launched a host of unusual whiskies, many of
which are bottled straight from the cask, and it is
also experimenting with peat levels and unusual
finishes. The popularity of Edradour as a visitor
attraction is remarkable, given that it is one of
Scotland's smallest distilleries. It makes just 12
casks of whisky a week, and getting hold of the
drink is not always easy. But Edradour commands
huge loyalty from those who have discovered it,
particularly if they have visited the distillery.

Scotland's Distilleries

Fettercairn HIGHLANDS

Laurencekirk, Kincardineshire
www.whyteandmackay.co.uk

Signature malt
Fettercairn "1824" 12-year-old

Fettercairn was licensed in 1824,
making it one of the oldest legal
distilleries in Scotland. But
it has had a mixed history, and to
this day has been something of a
misfit. The distillery itself is a
pleasing one, set in the most
rustic of environments and close
to the pretty Georgian village of the same name.

Taste-wise, Fettercairn isn't as weighty
as many of its Highland neighbours, and its
unassertiveness has meant that it has been
overlooked by many as a single malt. Most of the
output goes into Whyte & Mackay's blends,
where its unusual taste profile seems to excel.

Glenallachie SPEYSIDE

Aberlour, Banffshire

Glenallachie was founded in 1967 and has been a sizeable contributor to a range of blended whiskies ever since. It was the last distillery to be designed by the great distillery architect of the 20th century, William Delmé-Evans, who died in 2002.

It's a modern and functional distillery which uses water taken from a spring on nearby Ben Rinnes. The malt is mainly used in blends and its current lifeline was established in 1989 when it was taken over by Pernod Ricard. Single bottlings are rare. A cask strength version aged about 15 years was released in 2005.

The whisky itself is delicate and floral, a pleasant Speysider worth seeking out if you can.

Glenburgie SPEYSIDE

Forres, Morayshire

Glenburgie is a tale of two eras. The original distillery, dating from 1829, hit top gear in the late 1950s, when it was expanded to help meet the demand for malts to put into blended whisky. It housed two Lomond stills – tall pot stills with plates in the neck designed to alter the reflux of the still. However, the Lomond stills were very hard to maintain, and ceased to be used. Malt produced at this time occasionally appears under the name Glencraig.

The modern era began in 2005, after the distillery had been rebuilt at a cost of more than £4 million. Bottlings are very rare but worth seeking out. The whisky has gingery and dark chocolate characteristics, offering an unusual and pleasing experience.

Glencadam HIGHLANDS

Brechin, Angus
www.angusdundee.co.uk

Signature malt
Glencadam 15-year-old

When it closed in 2000, Glencadam looked like it had gone for good, but just three years later it was bought by Angus Dundee Distillers and brought back to life.

As under its previous ownership, most of Glencadam's whisky is destined for a range of blends, including Ballantine's, Teacher's and Stewart's Cream of the Barley.

However, Glencadam is something of a hidden gem, and the release of a 15-year-old from the distillery was welcomed in a number of quarters. That's not at all surprising, because it's an extremely drinkable and pleasant malt.

Glendronach SPEYSIDE

Forgue, near Huntly

Signature malt
Glendronach 12-year-old: rich in red berry fruit
flavours and intense malt

Special occasions
Glendronach 33-year-old: sherry cask perfection, with toffee, Crunchie bar and a mouth-watering wood and malt balance; stands up to its age

Shut for six years between 1996 and 2002, Glendronach's reprieve came in the form of demand for the Ballantine's blend, to which the distillery contributes some of the malt content. Under new ownership, there are big hopes for what is an underrated malt. The distillery is a traditional one – with wooden washbacks, almost exclusive use of ex-sherry casks and stone dunnage warehouses for maturation.

Glendullan SPEYSIDE

Dufftown, Keith, Banffshire
www.malts.com

Special occasions
All the Glendullan Rare Malts are worth
tasting, but they can be hard to find

The Glendullan Distillery can
produce 3.7 million litres a year,
yet it is virtually unknown as a
single malt whisky in the UK.
In the US, though, Diageo has
bottled the whisky under its
umbrella name "The Singleton",
and it is being warmly received.

In the UK, a 12-year-old Glendullan was
available for a while, and some is sold as an
8-year-old through supermarkets. Glendullan
was chosen as the Speaker's whisky by Betty
Boothroyd, Speaker of the House of Commons,
in 1992, and there was a special bottling to
celebrate the distillery's centenary in 1997.

Glen Elgin SPEYSIDE

Longmorn, Elgin, Morayshire
www.malts.com

Signature malt
Glen Elgin 12-year-old

Special occasions
Glen Elgin 32-year-old

Few distilleries have had a more
rocky existence and survived to
tell the tale. Opened at the
beginning of the 20th century,
just as the industry was falling in on itself,
Glen Elgin was closed and sold four times
in its first six years. Today, it is owned by the
drinks giant Diageo.

Glen Elgin attracts attention from enthusiasts
because it has six worm tubs for condensing the
spirit – a slow method that produces a characterful whisky. Besides a few special bottlings that
have been released, Glen Elgin is most closely
associated with the White Horse blend.

Glenfarclas SPEYSIDE

Ballindalloch, Banffshire
www.glenfarclas.co.uk

Core range
Glenfarclas 10-, 12-, 15-, 17-, 21-,
  25-, 30-, 40- and 50-year-olds
Glenfarclas 105 Cask Strength

Signature malt
Glenfarclas 12-year-old: more fruit
and oak than the 10-year-old, and
lashings of sweet malt

Special occasions
Glenfarclas 30-year-old: rich, fruit cake chewiness and
lots of chocolate and orange – wonderful

Despite the competitive demands of a global
market, there is still something wonderfully old-
fashioned about Glenfarclas. It eschews any form
of gimmickry, focusing instead on its strengths –
malts produced in top quality sherry casks. These
are robust whiskies that stand up well to ageing –
hence the 40- and 50-year-old expressions.

Glenfiddich SPEYSIDE

Dufftown, Banffshire
www.glenfiddich.com

Core range
Special Reserve 12-year-old
Caoran Reserve 12-year-old
Solera Reserve 15-year-old
Ancient Reserve 18-year-old
Glenfiddich 30-year-old

Signature malt
Special Reserve 12-year-old: no frills
fruity Speysider with the drinkability
factor turned up high

Special occasions
The rich, soft and lush chocolate flavours in the
30-year-old are worth seeking out

Glenfiddich is the single malt that lit the touch
paper to start the malt whisky explosion. It began
in the 1960s and, in the UK at least, "Glenfiddich"
soon became synonymous with "malt whisky".
Glenfiddich's owner, William Grant, was not only

the first company to promote a single malt, but was also the first to open up the secrets of the malt world by opening a visitor centre. No fan of malt whisky should ever forget that, and nor should this whisky be dismissed as a novice's whisky just because it has been around so long. It has maintained its position as the world's biggest selling malt for good reason.

Glenfiddich's owner has continued to invest in the distillery to make sure that it still has a home worthy of its world status, and everything here is stylish and impressive. And because Glenfiddich shares a site with the traditional and more "serious" malt distillery Balvenie, there is something here for both beginner and seasoned whisky enthusiast.

A brand new visitor centre, shop and restaurant ensure that Glenfiddich stays at the top of its game.

Glen Garioch HIGHLANDS

Meldrum, Aberdeenshire
www.glengarioch.com

Core range
Glen Garioch 8-, 15- and 21-year-olds

Signature malt
Glen Garioch 15-year-old: an excellent introduction to
Highland malt – a touch of oak and smoke around a
mass of malt, and with green fruit and a distinctive
Glen Garioch earthiness

Glen Garioch (pronounced
"Glen Geery") is a quirky
little distillery that
produces a whisky that's
hard to pin down. This is
partly because there are
peated and unpeated
versions, but there are also
distinctive notes to Glen
Garioch that are uniquely
its own.

Glengoyne <small>HIGHLANDS</small>

Drumgoyne, near Killearn
www.glengoyne.com

Core range
Glengoyne 10-, 12-, 17-, 21- and 28-year-olds

Signature malt
Glengoyne 10-year-old: clean, crisp fruity
malt that shows off all the distillery
characteristics

Special occasions
The 16-year-old Scottish Oak, if you can get it,
or the 21-year-old, which has a creamy quality and
some deep, fruity, almost blood orange, notes

Glengoyne's pretty, 200-year-old distillery is set
in a wooded area associated with Rob Roy, and it
is one of the few surviving distilleries in this part
of Scotland. Unchallenging, clean and pure
tasting, Glengoyne whiskies are easy to drink and
ideal for the novice. However, an extensive range
of older and vintage malts guarantees a challenge for the more experienced palate, too.

Glen Grant SPEYSIDE

Rothes, Morayshire

Core range
Glen Grant (no age)
Glen Grant 5-year-old
Glen Grant 10-year-old
Lord of the Isles

Signature malt
Glen Grant: clean and
crisp, like a green apple

Special occasions
Gordon & MacPhail have some Glen Grant that's been
aged for more than 30 years – well worth exploring!

Glen Grant sits in the heart of Speyside but has a
personality and charm like no other in the region.
The whisky itself is pale and young. Bottles of
Glen Grant are sold by the millions in mainland
Europe, particularly in Italy, but surprisingly little
elsewhere. The Italian drinks giant Campari
bought it in 2006.

Glenkinchie LOWLANDS

Pencaitland, Trenent, East Lothian
www.discovering-distilleries.com

Core range
Glenkinchie 12-year-old
Distillers' Edition 14-year-old

Signature malt
Glenkinchie 12-year-old: light
and easy with a hint of ginger

With Edinburgh just a bus ride away, Glenkinchie
is the nearest thing the Scottish capital has to its
own malt, and it would be all too easy to dismiss
this as a token malt – light, refined and suitably
easy on the palate to reflect the more genteel
face of Scotland. It is, however, a quality whisky,
with a dry, spicy finish.

Glenkinchie represents the Lowlands in
Diageo's original Classic Malts range. Its gentle
personality makes it an ideal aperitif whisky.

Glenlivet SPEYSIDE

Ballindalloch, Banffshire
www.theglenlivet.com

Core range
The Glenlivet 12-, 15-, 18 and
  21-year-olds
The Glenlivet XXV
Nadurra
1972 Cask Strength

Signature malt
The Glenlivet 18-year-old: not only
a signature malt but also classic
Speyside, with rich apple and
berry fruits, and clean, fresh malt

Special occasions
The French Oak Reserve is wonderful, but if you're at a
duty-free shop stocking the Nadurra 16-year-old, that's
the one. Look for the cask strength version, with
lashings of malt, chocolate and spice

The first licensed distillery in Scotland is also one
of its best. As a place to visit, as a producer of
exceptional whisky and in historical terms too,

Glenlivet has very few rivals on Speyside. The distillery's owner, Pernod Ricard, wants Glenlivet malt to challenge Glenfiddich for the number one spot worldwide, and the distillery is growing its core brands with this in mind. However, the company also has to satisfy the thirst of enthusiasts for rarer malts from the archives, making the distillery both commercial and esoteric. Glenlivet has even let its whisky veteran Jim Cryle have his own mini-still, so that twice a year he can distil spirit in the way that it would have been 200 years ago.

What a great combination: a distillery with superb whisky; a couple of enthusiastic eccentrics at the helm; and enough tales of derring-do in the distillery's history to fill a *Boy's Own* annual several times over.

Glenlossie SPEYSIDE

Elgin, Morayshire
www.malts.com

Signature malt
Flora & Fauna 10-year-old

Glenlossie sits next door to another distillery, Mannochmore, and shares the same workforce and warehouses. The distilleries are chalk and cheese, with Mannochmore built in the 1970s and Glenlossie established a whole century earlier. These days Glenlossie is part of the Diageo empire and is used mainly for blending, where it enjoys a strong reputation. It operates only between October and March, though its output is highly rated by whisky blenders.

Glenlossie is a rarity as a single malt but is highly regarded for its outstanding quality, so look out for any independent bottling. A limited edition 10-year-old was released in the early 90s as part of the Flora and Fauna series.

Whisky barrels at Glenmorangie (see next pages)

Glenmorangie HIGHLANDS

Tain, Ross-shire
www.glenmorangie.com

Core range
Glenmorangie Original
Nectar D'or
Lasanta
Quinta Ruban

Signature malt
Glenmorangie Original: complex spice
and oak dance around the malt with gay
abandon and thrilling effect

Special occasions
Limited edition Glenmorangie Margaux Cask Finish
Vintage 1987: whisky making at its finest, and a must
for any fan of this distillery; bursting with complex
flavours and vim

Glenmorangie may be one of the giants of
whisky, and therefore taken for granted by some,
but it is also among a handful that spare no
expense in sourcing the finest oak in which to
mature their whisky. The quality of the malt

here is going from strength to strength and, to highlight that fact, the distillery overhauled its range in late 2007.

The Nectar d'Or spends 10 years in ex-bourbon casks, followed by a spell in ex-Sauternes wine barriques. It is heavy on the flavour, but the lemon, grapefruit and spice are attractive enough. The Lasanta is a new look; new taste, but fine Glenmorangie with sherry wrapping and an enjoyable nuttiness. Quinta Ruban is matured in ex-bourbon casks, then finished in port pipes from wine estates – called quintas – in Portugal.

Some distilleries just drip with style and class, and this is one of them. If you're wanting to really spoil yourself, stay in a country house nearby and live like a laird for a while. When you explore the estate, first visit the Tarlogie spring that releases its precious mineral-rich water after a few hundred years permeating through rock, then marvel at the time and dedication the

The distillery at Glenmorangie

distillery puts into making its whisky. With its rugged coastline and bracing breezes, just being around Glenmorangie's distillery makes you feel healthy and vital, too. Though it means quite a long trek through the Scottish Highlands, the journey to this particular distillery is well worth the effort.

Glen Moray

Elgin, Speyside
www.glenmoray.com

Core range
Glen Moray Classic, 12- and 16-year-olds

Signature malt
Glen Moray 12-year-old: classic Speyside
whisky with fruit, honey and malt all in
balance – simple, but beautifully executed

Special occasions
The 1991 Mountain Oak Malt The Final
Release: spicy, warming and richly sweet,
with a hint of ginger

Scotland's Distilleries

Until late 2008 Glen Moray was little more than
a bit-part player, its malts widely used to service
the discount end of the whisky market. However,
LVMH has offloaded the distillery and its new
owners are seeking to give it a makeover. It won't
be easy, but the cause is helped by good quality
oak and some very good whisky, especially the
vintages and distiller's choices. One to watch.

Glen Ord HIGHLANDS

Muir of Ord, Ross-shire
www.discovering
distilleries.com

Signature malt
Glen Ord 12-year-old

Glen Ord is one of
owner Diageo's biggest
producing distilleries, but
it is something of a journeyman malt, and a
succession of name changes has done little
to help it build a reputation.

Diageo is targeting the Asian markets, where
The Macallan has long been dominant. To
compete, Diageo needed a sherried whisky to
rival Macallan's, so the sherry cask content in
Glen Ord has been upped, and the whisky
rebranded for Taiwan as The Singleton of Glen
Ord. With malt in short supply, allocation of
supplies to overseas territories has perhaps been
inevitable. Let's hope the trickle of malts going
down this route doesn't become a flood.

Glenrothes SPEYSIDE

Rothes, Aberlour
www.glenrotheswhisky.com

Core range
The Glenrothes Select Reserve
& Vintages

Signature malt
The Glenrothes Select Reserve:
honey, fruit and spices from perfect
oak casks – magnificent

Special occasions
Any Vintage of the 1970s: few distilleries
put out such consistently fine malts

The distillery is a big producer, providing whiskies
for several blends, including Cutty Sark. The
single malt is the epitome of sophistication
and style. It is packaged in distinctive, grenade-
shaped bottles with hand-written labels. The
whisky is excellent, and if you want Speyside fruit
and honey with the volume turned up to 11, this
is the distillery to seek out.

Glen Scotia HIGHLANDS

Glen Scotia
www.lochlomonddistillery.com

Core range
Glen Scotia 12-year-old
Glen Scotia 17-year-old

Signature malt
Glen Scotia 12-year-old

Campbeltown on the west coast of
Scotland used to be a rich and
vibrant whisky-producing region. It
saw no fewer than 34 distilleries set up here in its
19th-century heyday. Now just two are producing
whisky for the market, Glen Scotia and
Springbank, with four-year-old Glengyle set to
become the third. Now run by Loch Lomond
Distillery Company, Glen Scotia produces only
750,000 litres a year, making it one of Scotland's
smallest producers. Though a relatively rare malt,
if you can track down an independent bottling, it
will be something to treasure.

Glen Spey SPEYSIDE

Rothes, Morayshire
www.malts.com

Signature malt
Glen Spey Flora & Fauna 12-year-old

There is considerable debate among Speyside lovers as to which town is the spiritual capital of the region. Certainly Rothes, rich in history and blessed with four working distilleries, has a strong case to argue.

The least known of the four Rothes distilleries, Glen Spey is one of those Diageo Speyside workhorses that make malt primarily for the blended whisky market – in this case, particularly for J&B. Although single malt bottlings are rare, there have been a number of independent releases and, a few years ago, a 12-year-old was released in Diageo's Flora and Fauna range.

Glentauchers SPEYSIDE

Mulben, Keith, Banffshire

One of the anonymous but sizeable producing distilleries now owned by Pernod Ricard in Speyside, Glentauchers produces 3.4 million litres of spirit a year for inclusion in Ballantine's blended whisky. Pernod Ricard has ambitious plans for this internationally well-known blend, so Glentauchers' future would seem secure as one of its key malt suppliers. Single malt bottlings remain very rare, even though several whisky writers rave about the Glentauchers malt.

Glenturret HIGHLANDS

Crieff, Perthshire
www.famousgrouse.co.uk

Signature malt
10-year-old: rich, bold and honeyed,
with a strong malt backbone

Special occasions
The Whisky Exchange in London and
Douglas Laing have both released 27-
year-old independent bottlings

Glenturret has a fine pedigree. It
was founded in 1775 and may well be the
oldest working distillery in Scotland. It is small,
producing about 300,000 litres per year, and
most of the output goes into The Famous
Grouse, which is the best-selling blend in
Scotland. That doesn't leave much whisky for
single malt bottlings, but, unusually, the distillery
does offer vatted (blended) malts, aged at 12,
18 and 30 years.

Highland Park ISLANDS

Kirkwall, Orkney
www.highlandpark.co.uk

Core range
Highland Park 12-, 15-, 16-,
18-, 25-and 30-year-olds

Signature malt
Highland Park 12-year-old:
soft fruits wrapped in
honey and rounded off
with a gentle smokiness

Special occasions
The 18-year-old: the
trademark honey, malt and fruit are given an extra
dimension by the presence of wood, smoke and spice

Highland Park lies on rugged and weather-swept
Orkney, and its malt is equally hardy. But Orkney
is a warming, soulful place too, and the malt is
outstanding, combining fruit, honey, spice oak
and a degree of peat from barley malted on site.
HP has a large following, with good reason.

Inchgower SPEYSIDE

Buckie, Banffshire
www.malts.com

Signature malt
Inchgower Flora & Fauna 14-year-old: sweet and inoffensive Speyside malt, with a touch of earthiness, and even saltiness

Inchgower has the capacity to produce a sizeable amount of whisky – in excess of two million litres each year – though most of it is used for owner Diageo's heavyweight blends, including Bell's and Johnnie Walker. However, Diageo has released an Inchgower 22-year-old and 27-year-old as part of its Rare Malts series, both of which are excellent.

It's a pretty distillery, situated near the coast in the north of the Speyside region; the coastal proximity might explain why it's not a typically sweet Speysider.

Scotland's Distilleries

Jura ISLANDS

Craighouse, Isle of Jura
www.isleofjura.com

Core range
Jura 10-year-old
Jura 16-year-old Superstition
Jura 21-year-old

Signature malt
Jura 10-year-old: young and fresh
tasting with some melon in the
malt and a trace of smoke

Special occasions
Jura 21-year-old Cask Strength

Though it often seems to be in the shadow of the
whisky metropolis across the water on Islay, Jura
is a top-notch distillery and produces a very fine
malt in its own right. It also occasionally likes to
release a peated malt, just to prove it can bark as
loudly as its neighbours on Islay, if it so chooses.
A very young peated Jura is definitely worth
seeking out.

Kininvie SPEYSIDE

Dufftown, Moray

Kininvie is one of Scottish whisky's best-kept secrets. It is hidden away behind Glenfiddich and Balvenie distilleries, and, although its owner William Grant has talked about releasing a single malt from Kininvie, this has yet to happen. The distillery's purpose is to provide malt for blending, such as for William Grant's Monkey Shoulder blended malt whisky, and, so far, it has been fully employed in this pursuit. Perhaps, now that the company has opened a new malt distillery to ensure supplies in the future, there will be sufficient whisky at Kininvie for it to be bottled in its own right.

Knockando SPEYSIDE

Knockando, Aberlour, Banffshire
www.malts.com

Core range
Knockando 12-, 18- and 21-year-olds,
plus other vintages without age
statements

Although regarded as an elegant
and complex whisky, Knockando
has had only a very small presence
in the single malt market in the
UK. On the Continent and in
the US, however, it is more widely distributed.
Un-aged single malt bottlings do appear in the
UK from time to time, and in some markets older
expressions of Knockando are released. The malt
is also one of the key whiskies in the J&B blend.

Knockdhu HIGHLANDS

Knock, By Huntly,
Aberdeenshire

Core range
anCnoc 12-, 16 and
    30-year-olds
anCnoc 1991

Signature malt
anCnoc 12-year-old: a bit
of everything here – spice,
malt, oak and fruit, in
perfect balance

Special occasions
anCnoc 30-year-old: not for the faint-hearted, but
worth seeking out if you like big, bold, oaky whisky

The distillery is called Knockdhu, but, because
it was often confused with nearby Knockandu,
owner Inver House decided to call the single
malt by its Gaelic name, anCnoc ("the hill"). Its
complex, earthy and clean taste owes much
to traditional production methods.

Lagavulin <small>ISLAY</small>

Port Ellen, Islay
www.malts.com

Core range
Lagavulin 12-year-old
  Cask Strength
Lagavulin 16-year-old
Distiller's Edition Double Matured

Signature malt
Lagavulin 16-year-old: a
masterclass in peat working at
different levels

Special occasions
Lagavulin 12-year-old Cask Strength: it's becoming very
rare, but this super-charged and younger version of the
classic 16-year-old is an altogether feistier, fuller and
more challenging whisky

Along with Ardbeg and Laphroaig, Lagavulin
completes a "holy trinity" of distilleries in
southeast Islay that have perfected the smoky
and phenolic style of whisky for which the
island is famous.

Like Ardbeg and Laphroaig, Lagavulin has the sea lapping at its doorstep and is everything you'd hope that a distillery should be. Lagavulin's warehouses are among the most atmospheric you'll find anywhere in Scotland.

In recent years, Lagavulin has suffered some major stock shortages, and its absence has made many hearts grow all the fonder. No-one has ever doubted its quality, but the shortages have given it what can only be described as an iconic status.

Lagavulin 16-year-old is a true giant from the peated isle. A massive dose of peat on the nose; equally strong and smoky on the palate, with cocoa and liquorice, and a rich, deep, growling body. Stunning!

Laphroaig ISLAY

Port Ellen, Isle of Islay
www.laphroaig.com

Core range
Laphroaig 10-year-old
Laphroaig 10-year-old
  Cask Strength
Laphroaig 15-year-old
Laphroaig Quarter Cask

Signature malt
Laphroaig Quarter Cask: this offers new
whisky enthusiasts an opportunity to
experience the unique peat characteristics of
Laphroaig without the full flavour bombardment that
typifies the 10-year-old

Special occasions
Laphroaig 27-year-old: released in late 2007, this is a
rare and truly remarkable Laphroaig, with a weighty
wave of plummy sherry and bonfire smoke, before a
chunky peat and wood finish that goes on and on...

Laphroaig (pronounced "Laff-roy-g") is probably
the most iconic Islay brand – the Marmite of

whisky, which you either love or hate. Those who love it, tend to really, really love it; and for whom few other malts can compare.

First impressions of Laphroaig are that it's all smoke, fish and medicine. But spend some time with it, and there is an impressive array of flavours behind the Vesuvian-like sardine and smoke attack. As an entry level malt, try the Quarter Cask. This is a whisky finished in smaller casks, accelerating the maturation and softening the peat attack to great effect.

If you want to walk the tightrope without a safety net, then Scotch doesn't get much better than the 10-year-old Cask Strength. This is a massive peat attack, along with the richest malt and fruit double whammy found in any malt. The 27-year-old is another twist: all plums, stewed strawberries and liquorice in with the smoke, and a long finish. Also look out for special bottlings.

Linkwood SPEYSIDE

Elgin, Morayshire
www.malts.com

Signature malt
Linkwood 12-year-old: more floral than
fruity; wispy, subtle and rounded

Linkwood is one of the most attractive
and intriguing distilleries in Scotland.
The whisky is highly respected; the
distillery location, surrounded as it is
with a nature reserve, is quite
wonderful; and the strange production set-up
keeps the "trainspotters" in business for hours.

There are two sets of stills on site: one set
produces the bulk of the spirit, while an older set
is employed for some of the year to produce a
different spirit; the two are then mixed before
filling to cask. Linkwood's distinctive whisky is
particularly popular among blenders, while rare
bottlings of single malt, whether official or
through the independent sector, are much
sought after by a hardcore band of devotees.

Loch Lomond HIGHLANDS

Alexandria, Dumbartonshire
www.lochlomonddistillery.com

Loch Lomond is like no other distillery
in Scotland, and has more in common
with one of the large Irish or Canadian
distilleries, with pot stills, a grain plant
and "rectifiers" all employed to make a
range of different whisky styles, most
of which are used for the company's
own blends. The reason for producing
so many types is to help overcome
shortages of malt, a problem that may well be
reappearing for independent blenders as demand
for whisky rises. So Loch Lomond, Old Rhosdhu
(sometimes bottled as a surprisingly youthful 5-
year-old) and Inchmurrin all hail from here.

The many styles of whisky mean that there is
no recognisable style, and in fact single malts are
relatively rare. But when they have appeared they
have been of an impressively high standard.

Longmorn

Elgin, Morayshire

Signature malt
Longmorn 16-year-old: a rich,
complex and weighty malt; and, at
48% abv, a big hitter all round

Special occasions
Longmorn 17-year-old Distillers
Edition: a masterpiece at cask
strength – an oral pillow fight as
fruit, oak and barley all battle for
supremacy. Exceptional!

Longmorn is the whisky equivalent of a cult
French-language film – adored by aficionados of
the cinematic art; ignored in other quarters. Or,
at least, that was the case. Owner Pernod Ricard
seemed content to allow the whisky's reputation
to rest on the back of some outstanding
independent bottlings until a few years ago,
when a cask strength 17-year-old was released.
That has since been followed by an official
16-year-old release.

A cooper preparing casks for The Macallan (see next pages)

Macallan SPEYSIDE

Craigallachie, Moray
www.themacallan.com

Core range
The Macallan Fine Oak
  10-, 15-, 21- and 30-year-olds
The Macallan Sherry Oak
  10-, 12-year old
  Elegancia, 15-, 21-, 25-
  and 30-year-olds

Signature malt
The Macallan Fine Oak 15-
year-old: a mix of bourbon
and sherry cask whisky that is laced with cocoa,
orange and dried fruits, and lays bare the rich quality
of The Macallan's malt

Special occasions
The Macallan Sherry Oak 18-year-old: quite possibly the
perfect age for a Scotch single malt. The oak tempers
the sherry here, while spices, dried lemon, orange peel
and an underlying sweetness all combine to produce a
classic single malt

Famed for its attention to detail, its refusal to cut corners and for the quality of its sherried whiskies, The Macallan has long enjoyed a loyal and passionate following.

Macallan really became the complete package when it launched its Fine Oak range a few years ago. By combining sherry and bourbon casks, The Macallan has reined in the dominant winey notes and created a clean, fresh and sophisticated range of whiskies. The Fine Oak 15-year-old is the best expression of this range.

The distillery itself is beautiful, set high on an estate overlooking the Spey. At its centre is Easter Elchies House, now used to entertain guests. The still house is highly impressive, with small squat stills like beer-bellied penguins. The distillery is open to visitors most of the year: the standard tour includes a glass of 10-year-old, but you can opt for the "precious tour", which includes a tasting of five Macallan malts.

Macduff <small>SPEYSIDE</small>

Macduff, near Banff

Core range
Glen Deveron 10-
and 15-year-olds

Signature malt
Glen Deveron
10-year-old

Confusingly, the
small amount of
single malt
produced by
Macduff Distillery is bottled under the name Glen
Deveron, which alludes to the local river. The
distillery was opened in the 1960s to provide
blending stock, notably for William Lawson, but
the malt is worthy of investigation in its own
right because it is atypical of Speyside whiskies.

Mannochmore SPEYSIDE

By Elgin, Morayshire
www.malts.com

Signature malt
Mannochmore 12-year-old

Lying to the south of Elgin, Mannochmore was established in 1971 to help provide malt for the Haig blend during a boom time for whisky. Consequently, Mannochmore is a rare beast as a single malt.

The distillery is famed for having produced the "black whisky" Loch Dhu, a decidedly average whisky that is, nevertheless, still in demand among collectors. An empty bottle once sold on eBay for £80, and when independent retailer The Whisky Shop released some Loch Dhu from its vaults, the bottles were selling for £175.

The malt, if you can find it, is a no-nonsense, relatively delicate but pleasant Speysider.

Miltonduff SPEYSIDE

Elgin, Morayshire

Miltonduff is another of the great distilleries formerly owned by the Canadian whisky giant Hiram Walker. The distillery's purpose remains primarily to produce blending malts. It went through a period of using Lomond stills – which were designed to produce an array of different styles of malt from the same still. The experiment was abandoned because Lomond stills are inefficient and notoriously difficult to clean, but, while they operated, the whisky produced was known as Mosstowie, bottles of still appear from time to time. These days, Miltonduff has the capacity to produce more than five million litres of malt, and it is a key component of Ballantine's. The distillery was acquired in 2005 by Pernod Ricard, which has since made Miltonduff its trade and production headquarters.

Mortlach SPEYSIDE

Dufftown, Banffshire
www.malts.com

Signature malt
Mortlach 16-year-old: a flavour-rich, chunky, oily and quirky malt, which tastes like nothing else on Speyside

Whisky enthusiasts adore Mortlach. It has a complex and unique distillation process that includes a motley crew of stills and a partial triple distillation, which ensures that all sorts of compounds are kept in the mix to give the whisky a variety of subtle nuances unlike anything else in the region.

Blenders love Mortlach too, and it is widely considered to be an "adhesive malt" that can bring lots of other flavours to order. The distillery is sizeable, and capable of producing three million litres of spirit a year.

A very small amount of 32-year-old Mortlach was released a few years ago.

Oban

Oban, Argyll
www.malts.com

Core range
Oban 14-year-old
Oban 1980 Distillers Edition
  Double Matured
Oban 32-year-old

Signature malt
Oban 14-year-old: a growling,
purring vehicle that moves up the
gears from gentle start to rich,
fruity and reasonably smoky
monster; full and intriguing

Oban's distillery is in the heart of the pretty sea
port. It drips with character and charm, and still
uses worm tubs to cool the spirit from the stills,
making for a characterful final whisky. The
coastal location and the distinctive peatiness of
the malt makes it one of the finest Highland
distilleries to visit.

Old Pulteney HIGHLANDS

Wick, Caithness
www.oldpulteney.com

Core range
Old Pulteney 12-, 17- and 21-year-olds

Signature malt
Old Pulteney 12-year-old: outstanding
tangy, salty seaside character and
plenty of Highland bite; a rich whisky
and a very moreish one

Special occasions
The 17-year-old: citrus fruits, rich malt
and the trademark salt with some spice make this very
hard to resist; for a treat, this is not too pricey either

Pulteney is the UK's most northerly mainland
distillery. Under its current owner, Inver House,
the range has been repackaged. The distillery
often has special distillery-only bottlings available
too, which have been excellent. The single cask
offering takes some beating.

Port Charlotte

Bruichladdich,
Isle of Islay
www.bruichladdich.com

While there has been
talk of rebuilding a
distillery on the site
of the demolished
Lochindaal Distillery
two miles from
Bruichladdich,
the current crop of
wonderful Port Charlotte releases are distilled,
matured and bottled at Bruichladdich (see pages
62–3). But Port Charlotte warrants its own entry
because its whiskies are heavily peated, more in
keeping with the island style, and are very
different from the malts under the Bruichladdich
name. A new Port Charlotte has been put out
each year since 2005. The first was PC5 and the
one for 2008 is called PC8.

Collins Gem Whiskies

126

Royal Brackla SPEYSIDE

Cawdor, Nairn

Signature malt
Royal Brackla 10-year-old: a rich and rewarding sweet Speyside

Royal Brackla is situated in Cawdor, home of the castle that features in Shakespeare's *Macbeth*. It is approaching its 200th anniversary, having opened in 1812, and is one of only three distilleries able to use the word "Royal" – an honour granted because William IV was partial to this whisky. Most of the production goes for blending.

Scotland's Distilleries

Royal Lochnagar HIGHLANDS

Balmoral
www.malts.com

Signature malt
Royal Lochnagar 12-year-old

Special occasions
Selected Reserve Royal Lochnagar: a
limited edition release, often aged for
about 20 years

Royal Lochnagar is Diageo's smallest
distillery and one of its prettiest,
nestling on the edge of the Balmoral estate, the
Scottish home of the Royal Family. It is entitled
to use the prefix "royal" because Queen
Victoria visited the distillery in 1848 and took a
liking to it. It maintains old-fashioned equipment
such as wooden washbacks and "worms" – flat-
lying copper pipes for condensing spirit that pass
through a pool of cool water on the distillery roof.
Two small stills and a long fermentation period
contribute to a distinctive and weighty whisky.

Scapa ISLANDS

St Ola, Kirkwall, Orkney
www.scapamalt.com

Signature malt
Scapa 14-year-old: zesty, fresh and
moreish, with ripe melon, lemon and
honey flavours

Until a few years ago, Scapa was
an extremely rare whisky. The
distillery was all but abandoned,
producing spirit for only a few
weeks a year to top up supplies.
However, it was refurbished and put back into
production in 2003. Pernod Ricard then took it
over from Allied Domecq. As a result of the
changes, the standard 14-year-old is now more
readily available and worth seeking out.

Try some of Scapa's older and cask strength
expressions, which have a barley intensity and
some salt and peat notes that generate a
rewarding level of complexity.

Speyburn SPEYSIDE

Rothes, Speyside
www.inverhouse.com

Signature malt
Speyburn 10-year-old: clean
and simple, sweet and with
the faintest smoke undertow

Special occasions
Speyburn 25-year-old Solera Cask: not
an easy whisky to pin down because its
style is evolving. But what you can
expect is the age showing through, with
the Speyburn sweetness tempered by
spice and oak from the wood

Speyburn lies in the heart of Speyside. The
district is verdant and beautiful, and Speyburn's
pagoda-style chimneys make it an archetypal
Speysider. However, much of Speyburn's malt is
exported to America, and it is not particularly
known in its own right in Europe.

Speyside \text{SPEYSIDE} SPEYSIDE

Drumguish
www.speysidedistillery.co.uk

Core range
Drumguish
Speyside 8-, 10- and 12-year-olds

Signature malt
Speyside 12-year-old: the richest and
fullest of the distillery's malts

This is a neat and compact distillery,
a long way to the south of the area
most associated with Speyside but, nevertheless,
close to a major tributary of the Spey River. The
distillery was used in the filming of the *Monarch
of the Glen* TV series, but, in terms of whisky
making, it is something of a secret to many.

Speyside produces a range of malts, and the
company has its own Glasgow-based operation
for blending and bottling its whisky. It is not a big
player in the UK, and it's fair to assume that a
great deal of the distillery's whisky goes abroad.

Springbank HIGHLANDS

Campbeltown, Argyl
www.springbankdistillers.com

Core range
Springbank 10-year-old,
  15-year-old and 25-year-old
Springbank 10-year-old 100 Proof

Signature malt
Springbank 10-year-old, 100 Proof:
a full malt like no other – blatant and
colourful, yet nuanced, unpredictable
and engaging

Springbank is a "three whiskies" distillery. In
addition to the eponymous malt, it also produces
Longrow, a significantly peated whisky, and
Hazelburn, which is triple distilled. And, since
2004, it has also had a sibling called Glengyle –
the first new distillery in Campbeltown for 125
years; it produces a malt called Kilkerran.
Production at Springbank was cut back in late
2008 due to the high cost of oil and ingredients,
but the distillery's future is not under threat.

Strathisla SPEYSIDE

Keith, Banffshire
www.chivas.com

Core range
Strathisla 12-year-old
Strathisla 18-year-old

Signature malt
Strathisla 12-year-old: rich, sherried
and satisfying, with a nice platform
of sweet fruits

Special occasions
Strathisla 15-year-old Cask Strength:
bolder, oakier and arguably drier than the standard
bottling; the extra strength gives it added depth

Strathisla is the oldest working distillery in
the Highlands, and its maturation warehouse
contains some of Chivas Brothers' oldest stock,
along with rare and special casks, including one
that is owned by Prince Charles. Strathisla is a
fine whisky and it plays a key role in the outstanding Chivas blend.

Strathmill SPEYSIDE

Keith, Banffshire
www.malts.com

Signature malt
Strathmill 12-year-old: Released as part of the Flora and
Fauna series, Strathmill is sweet, honeyed and floral,
with a hint of orange

Strathmill is one of those classic and traditional
distilleries that ticks all the boxes when it comes
to the romance of malt. It is a pretty distillery,
with twin pagoda chimneys, situated by the side
of a river in the town of Keith, the epicentre of
Speyside.

It's a sizeable distillery, which includes in its
production process a purifier that's designed to
create a light style of whisky, much in demand
for blending – particularly for J&B. A 12-year-old
single malt was released for the first time in
2001, as part of Diageo's Flora and Fauna range.

Talisker distillery on the Isle of Skye (see next pages)

Talisker <small>ISLANDS</small>

Carbost, Isle of Ske

Core range
Talisker 10-year-old
1986 Distillers Edition Double Matured
Talisker 18-year-old
Talisker 25-year-old
Talisker 57 Degrees North

Signature malt
Talisker 10-year-old: classic pepper and
smoke explosion; a dry storm of a whisky

Special occasions
Talisker 18-year-old: this has everything
– lots of smoke, the trademark pepper and spice, a
honeycomb heart and a three-dimensional, chunky
depth not present in the 10-year-old. It is whisky at its
most wonderful

Nestling among the rocky crags and rugged
shoreline of Skye, Talisker is in perfect harmony
with its desolate surroundings. So too is its
whisky, which reflects the wild, stark landscape.
Skye is a rugged and unforgiving island, which

has witnessed some of the country's bloodiest and most dramatic history. Traditionally, its climate has been harsh and challenging. Unsurprisingly, then, there's an earthiness about the people and the place. But there's an other-worldliness to Skye, too, as if you have been transported to another planet. Both the distillery and the whisky echo this environment, and Talisker is a bold and confident malt – and decidedly masculine.

Talisker 18-year-old was voted the best malt in the world in the first World Whisky Awards, organised by *Whisky Magazine* in 2007, and with good reason. The trademark Talisker pepper and fire remains in place, but the age gives it a sweeter third dimension – faultless.

Tamdhu SPEYSIDE

Knockando, Aberlour

Tamdhu is the least known of Edrington's working distilleries – the others being Highland Park, Macallan, Glenrothes and Glenturret. Tamdhu has a major role to play, however, and, with the capacity to produce more than four million litres of spirit per year, it is a key contributor to blends such as The Famous Grouse. It is the only distillery still using "saladin boxes", a commercial method for malting barley. Tamdhu can also provide malt for the group's other distilleries.

The malt is relatively hard to find but it has an appealing fruit sherbet taste, and a clean and soft delivery on the palate.

Tamnavulin SPEYSIDE

Ballindalloch, Banffshire

Core range
Tamnavulin 12 Year Old

When Indian businessman Vijay
Mallya held a press conference to
announce his purchase of Whyte &
Mackay, he pulled the rabbit out of
the hat by announcing the reopening
of Tamnavulin Distillery after 12 non-
productive years. In 2007 the output
was a modest one million litres but in
2008 the distillery reached full production of
about four million litres, doubling Whyte &
Mackay's output.

Tamnavulin was built in 1966 – a newish
distillery – and is functional rather than pretty,
though it does lie in the scenic heart of the
Livet Valley, on a tributary to the River Livet.
Little of the previous production was ever bottled
as a single malt.

Scotland's Distilleries

Teaninich HIGHLANDS

Alness, Ross-Shire
www.malts.com

Signature malt
Teaninich 10-year-old

Global drinks corporation Diageo
has a number of Speyside distilleries
that operate in the shadows, and
none more so than Teaninich. Close
to the relatively famous Dalmore,
Teaninich is a sizeable distillery
capable of producing more than
2.5 million litres per annum, yet it is virtually
unknown in its own right.

The whisky has a couple of quirky production
characteristics that are of interest to the
technically-minded, but, for the most part,
Teaninich slips under the radar. A bottling did
appear in Diageo's Flora and Fauna range some
years back, but otherwise, as a single malt,
Teaninich is very much a rarity.

Tobermory

Tobermory, Isle of Mull
www.burnstewartdistillers.com

Core range
Tobermory 10- and 15-year-olds, Ledaig 10-year-old

Signature malt
Tobermory 10-year-old: light and refreshing, with
a blemish-free hit of malt through its heart

Tobermory produces a 10-year-old single malt,
which has to be matured on the mainland
because the distillery's warehouses were sold

off and converted into
flats. The distillery also
produces a peated
version, which goes by
the name of Ledaig
(pronounced "Led-chig").
It is starting to build a
powerful name for itself
and, unusually for a malt,
is very good when young.

Tomatin HIGHLANDS

Tomatin, Inverness-Shire
www.tomatin.com

Signature malt
Tomatin 12-year-old: a balanced and
easy-going, yet full Highland whisky

In 1974, Tomatin was the biggest
producer in Scotland. It still has
the capacity to produce a
significant five million litres of
spirit each year, yet Tomatin is a
strange beast, and not as well
known as perhaps it might be. The main reason
for this is because most of its whisky goes
abroad, either as single malt or through a
number of blends, most notably The Antiquary.
Occasionally, the distillery bottles vintage
expressions, which are worth seeking out. In
the past few years, there have been bottlings
from 1973 and 1965.

In 1986, Tomatin became the first Scottish
distillery to come under Japanese ownership.

Tomintoul SPEYSIDE

Ballindalloch, Banffshire
www.tomintouldistillery.com

Core range
Tomintoul 10-, 16- and 27-year-olds

Signature malt
Tomintoul 16-year-old: arguably the best value-for-money single malt in Speyside, commanding a modest price, but rich in fruit, malt and oak – stunning

Tomintoul can produce more than three million litres of spirit each year, and, since the turn of the millennium, it has been owned by independent bottler Angus Dundee. In that time, Tomintoul has been quietly building up its reputation as a single malt, and the 16-year-old is particularly impressive. The distillery has also launched a peated whisky called Old Ballantruan, and, along with BenRiach, is seriously challenging some preconceptions about the region.

Tormore SPEYSIDE

Advie by Grantown-on-Spey,
Morayshire
www.tormore.com

Signature malt
Tormore 12-year-old: easy-drinking,
clean and soft

Designed to be a showcase
distillery, Tormore is among
Scotland's quirkiest distilleries,
with oddball features such as a
musical clock. In recent years it
has become another sizeable
producer of malts intended for blends such as
Ballantine's and Teacher's. Pernod Ricard now
owns the distillery and has launched a delightful
12-year-old single malt, suggesting that the
distillery might be set for a spot in the limelight.

Tullibardine HIGHLANDS

Blackford, Perthshire
www.tullibardine.com

Signature malt
Tullibardine 10-year-old

Tullibardine had been mothballed for nine years when a consortium came together to buy the facility and its stocks in 2003. The first releases since then have come from the archives and include some whiskies that are more than 30 years old. The marketing of the whisky shows a modern approach, coupled with a strong emphasis on heritage (the shop and café are called 1488, to highlight the fact that beer was brewed on the site more than 500 years ago). The distillery's owners have not been afraid to try new ideas, and the policy seems to have paid off. But, as an independent, Tullibardine has been vulnerable to the threat of a takeover, and, in late 2008, the company was said to be considering its options for the future.

Openings and Closures

Whisky production has always ebbed and flowed in response to periods of high and low demand, but the current whisky boom, fuelled by huge interest from India, China and other parts of Asia, is unprecedented. As a result, there has been a frantic rush to up production and to make use of any spare capacity, resulting in the reopening of some mothballed distilleries and the construction of entirely new ones.

Recently reopened **Glenglassaugh** had been shut for more than 20 years. The new owners have responded to the excitement generated by the reopening to launch some very rare and old whisky distilled before the closure in 1986.

Braeval, which opened only in 1973, was originally known as the Braes of Glenlivet. It closed in 2002 but reopened in 2007, probably to help meet the demand for whisky for key Chivas blends such as Chivas Regal and 100 Pipers.

Both William Grant and Diageo are increasing their levels of production with new distilleries.

The former at **Girvan**, where it already had a grain plant and huge warehouse space; the latter with a super distillery at **Roseisle**, North Speyside. William Grant's is already in production and will produce high quality malt for blending purposes, while Diageo's is expected to come on line in 2009.

There's another new distillery in Speyside too. The **Huntly Distillery** is owned by Euan Shand, of independent bottlers Duncan Taylor, which is based in Huntly. In the Lowlands, **Daftmill** has been distilling since 2004, so its first bona fide whisky came to fruition in 2007 – though it's yet to achieve full maturation for release. It's a similar case with **Glengyle** in Campbeltown, and **Kilchoman** on Islay. All should be bottling their first whisky in the coming years.

Closed distilleries

Here is a round-up of some of Scotland's closed or mothballed distilleries. Bottlings of their whisky do still appear, though some are exceedingly rare.

- **Banff** (Speyside)

Closed in 1983; demolished in 1985. Some casks have been released by independent bottlers, and

Diageo still has stocks, but Banff was never commonly available as a single malt.

• **Brora** (Highlands)

Closed in 1983 but still standing. Different expressions have been released regularly in the last 10 years, and the Brora 30-year-old is one of the world's truly great malts.

• **Caperdonich** (Speyside)

Mothballed in 2002, but available officially as a cask strength 16-year-old. There have been various independent bottlings too.

• **Coleburn** (Speyside)

Closed in 1985. Very hard to find, though a release in Diageo's Rare Malts series in 2000 can be found at very good retailers.

• **Convalmore** (Speyside)

Mothballed in 1985. Independent bottlings and the occasional release from Diageo mean that it can be bought from about £50 upwards.

• **Dallas Dhu** (Speyside)

Closed in 1983, sold to Historic Scotland and now a distillery museum. Can be found through a number of independent bottlers.

• **Glen Albyn** (Highlands)

Stopped producing in 1983; demolished in 1986.

Released as part of the Rare Malts series in 2002, and there are some independent bottlings.

- **Glenesk/Hillside** (Highlands)

Closed in 1985. Some independent bottlings and the occasional release of both Hillside and Glenesk in Diageo's Rare Malts series.

- **Glen Keith** (Speyside)

Mothballed in 2000 but not too rare a whisky still. It is rumoured that the Craigduff and Glenisla releases from independent bottler Signatory come from Glen Keith.

- **Glenlochy** (Highlands)

Closed in 1983; demolished in 1992. It has since been released in Diageo's Rare Malts series and through some independent bottlers.

- **Glen Mhor** (Highlands)

Closed 1983; demolished 1986. There have been Rare Malts releases and independent bottlings.

- **Glenugie** (Highlands)

Closed in 1983. Some independent bottlings do exist, but they are few and far between.

- **Glenury Royal** (Highlands)

Increasingly difficult to find, though top retailers still have some. The Whisky Exchange has old and rare bottles, if you've got the budget to match.

Port Ellen malt

- **Imperial** (Speyside)
Mothballed in 1998; official releases from Pernod Ricard augmented by independent bottlings.
  - **Inverleven** (Lowlands)
  Some independent releases.
  - **Kinclaith** (Lowlands)
  Rarest of the rare. Closed in 1975, after less than 20 working years. You'll do well to find any of this under £600, and you can expect to pay in excess of £1000.
  - **Ladyburn** (Lowlands)
  Lasted less than 10 years before shutting in 1975. Just about unobtainable as Ladyburn, but Royal Mile Whiskies had a bottle called Ayrshire, which is thought to have come from this distillery.
- **Littlemill** (Lowlands)
Production stopped in 1992 and the buildings were mostly demolished four years later. Available as an 8- and a 12-year-old.
- **Lochside** (Highlands)
Mothballed in 1992. Official bottlings very rare and expensive; independent bottles at good retailers.

- **Millburn** (Highlands)
Production ceased in 1985, and much of the equipment ended up at Benromach. Millburn has been released in the Rare Malts series, and there is a small amount from independent bottlers.
- **North Port/Brechin** (Highlands)
Mothballed in 1983 and demolished 10 years later. You'll find some independent bottlings priced at around £50.
- **Pittyvaich** (Speyside)
Mothballed in 1993, officially closed in 2002, and increasingly difficult to find.
- **Port Ellen** (Islay)
Closed in 1987; arguably the most sought after of all closed distillery malts. You can get it if you're prepared to pay the high prices demanded.
- **Rosebank** (Lowlands)
Closed in 1993. Rosebank has attracted the attention of independent bottlers, so it is obtainable.
- **St Magdalene** (Lowlands)
Closed in 1983. Appeared in Diageo's Rare Malts series; some independent bottlings are in circulation too.

Rosebank malt

Visitor Attractions

Many of Scotland's distilleries welcome visitors, and you can find out about tours and booking via their individual websites. In addition, there are several other whisky visitor attractions, from museums in former distilleries to the Scotch Whisky Heritage Centre in Edinburgh.

Dallas Dhu

Forres, Morayshire ● 01309 676548
www.historic-scotland.gov.uk

Dallas Dhu is a closed distillery lying in the heart of Speyside. It was purchased by Historic Scotland and has been maintained as a visitor attraction. As no whisky is produced here, it operates under a different set of health and safety rules to those that apply to working distilleries, and this means that visitors can get closer to the machinery involved in whisky production, which is what the whole experience is geared towards. Dallas Dhu is open all year round, but with reduced hours during winter.

Dewar's World of Whisky at Aberfeldy Distillery

Dewar's World Of Whisky
Aberfeldy Distillery, Aberfeldy, Perthshire
01887 822010 • www.dewarswow.com

Dewar's is one of the world's most established blended whiskies, and Dewar's World of Whisky has become one of Scotland's top tourist sites. Based at the Aberfeldy Distillery, it operates as a separate facility to the distillery and is a modern and interactive family experience. It tells the story of the Dewar family and explains the skills of the whisky blender. Tours range from a rudimentary one to the stylish Signature Tour, which includes a tasting of five whiskies, taking in a 21-year-old Aberfeldy and the highly expensive world-class blend Dewar's Signature.

The Famous Grouse Experience

Glenturret Distillery, Crieff, Perthshire • 01764 656565
www.famousgrouse.co.uk

The Famous Grouse Experience has helped to ensure that Glenturret is the most visited distillery in all of Scotland. It offers an enticing combination of quaint old distillery and state-of-the-art, interactive visitor attraction. The three-dimensional projection room is great, with images that glide across the floor and walls. At one point, the floor even appears to be a frozen river, complete with cracking ice!

The Famous Grouse brand

The Museum of Malt

Glenkinchie Distillery,
Pencaitland, East Lothian
01875 342005 • www.dis-
covering-distilleries.com

Glenkinchie's stills

Situated to the east of
Edinburgh on the Firth
of Forth, Glenkinchie
Distillery launched its
Museum of Malt some
40 years ago, making it something of a pioneer
in terms of whisky tourism. Today, the exhibits
include a beautiful distillery model made 80
years ago and an impressive array of distillery
tools and equipment.

The Scotch Whisky Heritage Centre

354 Castlehill, The Royal Mile, Edinburgh
0131 220 0441 • www.whisky-heritage.co.uk

Situated at the top of the Royal Mile, near
Edinburgh Castle, this is another interactive
attraction. It's recently been renovated, though
the tour still explains the history of Scotch whisky
and the process by which it is made. And, of
course, there's a dram to be had too.

Blended Scotch

Despite popular belief, blends are not the inferior relative of malts. In fact, at their best blended whiskies demonstrate the full range of the whisky maker's skills, being far more complex and structured than a single malt can ever be. If a malt whisky is a solo musician, the blended whisky is an orchestra, skilfully balanced to produce a harmonious oral symphony. So why the poor reputation?

Part of the reason lies in the fact that a large proportion of the millions of bottles sold annually – still more than nine out of 10 whiskies sold are blends – are, indeed, of low quality. Another is that, compared to single malts, there is little dynamism in the sector because blends are made for consistency and new launches are rare.

But there are many very good blends even in the standard market, and at the premium end some of the whisky world's very best products fall in this category. Indeed premium blends could be set for a comeback.

Glenturret Distillery is the home of The Famous Grouse blend.

Anatomy of a blend

A blended whisky is one that contains a number of single malts mixed with whisky made from another grain. Malt whisky is made in batches using pot stills; grain whisky is made using a continuous process using a column or Coffey still in a manner similar to the way vodka or gin spirit is most often made.

Grain was originally added to malt to make it less harsh and easier to drink. Part of the problem that some blends have, to the detriment of the category as a whole, is that the proportion of grain to malt is high, and the quality of the grain used is low.

When it's done right, though, a blend combines all the varied flavours of the malt and

Blended Scotch

157

brings it all together through the grain. A good blend is like a great painting, with scores of different colours and hues brought together to create a work of art.

Just as with single malts, blends may have an age statement on the label, and this refers to the youngest whisky in the mix, including the grain. And occasionally something very special is launched, such as last year's The Last Drop – a blend containing 70 malts and 12 grains, all of them distilled before 1960 and therefore at least 48 years old.

The great blends
As a rule of thumb, if you're thinking of buying a blend, go for one of the best-known brand names – they sell well for a reason. The leading players include:

• **The Antiquary**
High Speyside and Highland malt content and outstanding quality.
• **Bailie Nichol Jarvie**
Another blend that is rich in malt content – mainly from Glenmorangie.

• Ballantine's
Has a large range of fine and some-times rare blends in its portfolio, none of which are less than excellent.

• Bell's Original
Last year saw the launch of a new version of Bell's, an unaged whisky replacing the 8-year-old.

• Black Bottle
This is arguably the best-value-for-money whisky on the market. It is unusual in that it is an Islay-influenced blend, so chunky smoke and peat compensate for a loss of balance.

Black Bottle

• Chivas Regal
A wonderful new 25-year-old has enhanced an already outstanding blend name. Chivas makes great use of excellent Speyside malt.

• Cutty Sark
Clean, crisp and perfectly weighted. Cutty Sark is light and refreshing in its young expressions; big and great in older versions.

Chivas Regal 25-year-old

Johnnie Walker
Black Label

- **Dewar's**
  Blending at its finest. The standard versions are balanced and tasty and, at the top end, Dewar's Signature is an absolute masterpiece by any whisky standards.
- **Famous Grouse**
  Top quality mainstream blend - not surprising when you consider The Macallan and Highland Park are at its core – sherry and citrus in the mix.
- **Grant's**
  Based around core malt Glenfiddich, this is a big family of impressive blends.
- **Hankey Bannister**
  Three expressions, all of them special, each of them a study in subtlety.
- **J&B**
  More than 40 malts are used in this Diageo monster, and the quality is there for anyone to taste. One of the world's biggest players.
- **Johnnie Walker**
  The blender's blend and outstanding at all levels, from Red and Black to Blue and Gold.

• Royal Salute
Proof, were it needed, that blends can be every bit as premium as single malts. Different versions – all special, most very expensive.

• Teacher's
The Billy Bremner of whisky – fully committed, rough round the edges, but honest and loveable and always worthy of selection.

• Whyte & Mackay
Impressive array of blends from standard shelf to a 40-year-old with 70 per cent malt content.

Royal Salute
21-year-old

Blended malt whiskies (vatted malts)
This category comprises whiskies in which several malts are mixed together *without* any grain whisky. Traditionally such whiskies were known as vatted malts, but now, somewhat confusingly, they are known as blended malt whiskies. Blended malts may contain only two malts but often are a mix of many more. They offer unique flavour combinations and a full flavoured but complex mix to savour.

One company, the inspirational and constantly delightful Compass Box, has also invented a new whisky category in this area – boutique whiskies. Among the very best blended malts are:

• Clan Denny Islay
A peated Islay greatest hits, beautifully balanced and thoroughly enjoyable. Its sister from Speyside is pretty good too.

• Compass Box Juveniles
Malt's answer to a children's ballet class: young, exuberant, enthusiastic, sweet and reasonably graceful to boot.

• Compass Box Oak Cross
Boundary-breaking whisky making. The company's Spice Tree, wonderful as it was, tested whisky laws to the full and it was eventually replaced by Oak Cross. This is a stunning whisky, mixing quality used wood with virgin French oak heads to give the whiskies a distinctive spiciness.

• Compass Box Flaming Heart
Smoke-wrapped tinned strawberries with some oak and pear thrown in for good measure. Part of the Limited Release range, this is unlike any other whisky and a sublime treat.

- **Johnnie Walker Green Label**
As fresh and clean as a healthy stroll
up a Speyside hill in the heart of winter,
this is a blemish-free combination of
quality malt.

- **Monkey Shoulder**
Contains just three malts and is a total
delight. It's young and zingy, with
grapefruit and other yellow fruits
zipping across the palate. Great
packaging too. One route to follow
as whisky moves forward.

- **Serendipity**
Named after a happy accident in
which young Glen Moray malt was
mixed with some very old and rare
Ardbeg whisky from Islay, depriving the Ardbeg
malt of its age or its status as a single malt.
Thankfully the accident produced a gem of a
whisky. Very hard to get hold of now, though.

- **Sheep Dip**
Another example of quality malts perfectly
balanced. Whyte & Mackay's Richard Paterson
is the master blender responsible. Sheep Dip's
brother blend is called Pig's Nose.

Compass Box:
Flaming Heart

Blended Scotch

Irish Whiskey

Scotland might have established itself as the most famous whisky-producing nation on the planet, but it wasn't always so. The Irish have a long and proud tradition of whiskey-making (note the addition of the letter "e" for the Irish version of the drink), and over a drop or two they'll tell you that they invented the original drink (*uisce beatha* in Gaelic) and exported it to the Scots, who proceeded to make it in the "wrong" way.

What is Irish whiskey?

It's commonly thought that the main difference between Irish whiskey and Scotch is the fact that Irish whiskey is distilled three times, and single malt only twice. This is partly true, but there are distilleries in Ireland that distil just twice, and there are a couple of Scottish distilleries that do so three times. Another difference often cited is that Irish barley is never kilned with peat smoke, unlike Scottish single malt. Not true: there are plenty of Scottish malts that have no peated barley in them, and there are a few Irish whiskeys that are made using peat. The third distinction often talked about is the uniqueness of Irish pot still whiskey. This is true, in so much as pot still whiskey is a unique style, although now, sadly, only a fraction of Ireland's output is traditional pot still whiskey.

As with Scotch, there are four styles of Irish whiskey. Ireland produces a small amount of malt whiskey, which is produced in the same way as in Scotland; a very small amount of it is peated. There is a tiny amount of single grain

left to right: Redbreast pure pot still whiskey; Bushmills single malt; Greenore single grain whiskey; Jameson blended Irish whiskey

Bushmills Distillery in County Antrim

whiskey also, triple distilled like much of the rest of Ireland's whiskeys. But there are two other whiskeys that set Ireland apart.

Pot still whiskey is unique to Ireland and is a whiskey made from a "beer", or wash, made up of malted barley and another grain – normally unmalted barley, but occasionally wheat. This mix produces an oilier, earthier style of spirit. Pot still whiskey is almost always triple distilled to produce a spirit that is smoother and rounder than Scotch, because the extra spell of interaction between spirit and copper removes a higher proportion of congeners.

The fourth style of Irish whiskey, and the most famous, is blended Irish, a mix of pot still, grain and occasionally single malt. Jameson, Black Bush and Powers are the most famous blends.

Ireland's distilleries

Ireland has only three major distilleries (plus one micro-distillery); together they produce about 30 different brands. The bad news is that you can't visit two of the main distilleries – a pity because they're the plants where the art of Irish whiskey making is best demonstrated. The exception is Bushmills, a beautiful and friendly distillery in Antrim on Northern Ireland's rugged coastline. It's the nearest Ireland gets to making whiskey like the Scots, because here they make single malt.

Bushmills

Just a few miles from the Giant's Causeway, a spectacular part of the British Isles, stands Bushmills. Like a once-great football team that still commands a large and loyal following but has drifted from the Premier League and never quite found its way back, Bushmills is doing okay but really ought to be doing a whole lot better.

However, it was sold to drinks giant Diageo a few years ago, so perhaps it will now get the support and exposure it deserves. And there are already signs that its malt and its excellent blend, Black Bush, are being promoted more energetically.

Bushmills single malt is triple distilled. This immediately puts it at a disadvantage economically, because its distilling costs are 50 per cent higher than those of its cousins just a short distance to the east in Scotland. High energy prices make that difference considerable.

But Bushmills doesn't shirk its responsibilities when it comes to quality, and the distillery also invests in the finest casks, including some outstanding sherry casks. If you're in the mood to spend a bit, go for the wonderful Bushmills 16-year-old. If not, try Black Bush, a blend with about 80 per cent malt content and a wonderful example of the art of Irish blending at its very best.

Midleton

A few miles away from Cork, in the south of the Republic of Ireland, is Midleton. It is the main producer of Irish whiskey and the centre for Irish Distillers, the owner of most Irish brands.

Midleton is one of the world's most impressive distilleries, though it's actually two distilleries on the same site.

The old distillery – now a visitor centre and re-christened the Jameson Heritage Centre – is kept in immaculate condition, so that you can all but see the workers who would have bustled through its spacious rooms before it closed down some 30 years ago. The new distillery, hidden away behind trees, is the whiskey equivalent of one of those vast industrial bases that James Bond stumbles upon when trying to stop Blofeld destroying the world.

Midleton Very Rare, one of Ireland's most expensive blends

Here, column stills and pot stills sit side by side on a massive scale and in high-tech splendour.

By employing a variety of combinations between the two distillation methods (continuous distillation in the column stills and batches of malt distilled in the pot stills), Midleton produces more than 25 different whiskeys. Every production variable is tweaked to make a diverse range of whiskey styles for different brands.

Irish Whiskey

Cooley

The stills at Cooley Distillery

Cooley is an independent distillery, situated north of Dublin on the Cooley Peninusla, not far from Northern Ireland. The distillery operates from an ugly former industrial alcohol plant, but Cooley has been an exciting addition to the Irish scene. It has resurrected some traditional whiskey names and is experimenting with a range of styles. Particularly noteworthy are its Connemara 12-year-old and Connemara Cask Strength whiskeys, both of which are peated single malts.

Kilbeggan

The village of Kilbeggan used to be home to Locke's Distillery. That distillery fell silent in 1953, but Cooley began leasing its warehouses for maturing whiskey a few years ago, and, as part

The former Locke's Distillery in Kilbeggan closed in the 1950s, but is now producing whiskey once more as Kilbeggan Distillery.

of the celebrations for Locke's 250th anniversary, Cooley began distilling again at Kilbeggan. Only the second distillation takes place on site at present, with the low wines being brought over from the main Cooley Distillery, but there are plans to put in a wash still in the near future. In the meantime, Kilbeggan produces about 325 litres of spirit a day.

American Whiskey

• •

The most well-known American whiskey style is bourbon (pronounced "ber-bun"). However, there are distilleries throughout the United States that make whiskey in other styles and employ different methods, including those that use malted barley and distil their spirit in pot stills, much as things are done in Scotland. Moreover, the most famous American whiskey, and one of the biggest selling spirits on the planet, Jack Daniel's, isn't a bourbon. It comes under the designation of "Tennessee whiskey". Jim Beam, on the other hand, is a bourbon, but in its most basic form, the standard white label version, it is an adequate but unexceptional representation of the style.

Tennessee whiskey

Jack Daniel's, the most notable non-bourbon whiskey, serves to illustrate how strict the rules are governing bourbon production. Jack Daniel's is made in Tennessee in pretty much the exact manner as bourbon whiskey. However, before it is put into the barrel after distillation, it is passed through a wall of maple wood charcoal. This procedure, known as the Lincoln County Process, helps remove fats and congeners, and makes the spirit smoother. But it is strictly forbidden under the rules governing the production of bourbon. So, like the other

Jack Daniel's Tennesse whiskey and Jim Beam Kentucky bourbon

Tennessee distillery, that of George Dickel, Jack is excluded. Quite happily, we might add, because down in Tennessee they're pretty confident that they are making a better product anyway.

What defines bourbon?

A few hours north of the Jack Daniel's home in Lynchburg is the state of Kentucky. While you don't have to come from Kentucky to make bourbon, it helps. America's greatest distilling names are grouped in an area that extends around the state capital of Louisville and stretches down to bourbon's capital, Bardstown. The distilleries based here include Jim Beam, Wild Turkey, Heaven Hill, Buffalo Trace, Maker's Mark, Woodford Reserve, Barton Brands and Four Roses. Kentucky casts a huge shadow over the rest of America's whiskey producers, as does bourbon.

Making bourbon

At its most simplistic, the process of making bourbon mirrors that of making single malt Scotch. You start with grain, make a wash, ferment it with yeast, distil it and then mature the resulting spirit in oak barrels. In practice, though, bourbon is bound by different rules

Modern fermenting tanks at Brown-Forman Distillery and traditional wooden vats at Woodford Reserve

and produced in a different way. Single malt Scotch is made using just one grain – malted barley. Bourbon uses a mix of grains, the dominant one being corn – 51% of the mash bill by law. Normally the mix of grains will be made up with two others, one of which is usually malted barley, as it is the best grain for providing the base for the conversion of

Woodford Reserve still uses pot stills for batch distilling.

sugars and enzymes into alcohol. Unmalted barley, wheat and rye are also used.

Distilling bourbon

The processes of creating a grist (the ground grain), fermentation and making the wash are similar to those for producing Scotch malt whisky. But the distillation process in most bourbon-producing distilleries is different.

American Whiskey

175

Instead of a batch of wash being distilled in a pot still, the wash is more commonly produced continuously, with distillation taking place in a column still, where the wash is forced down over steam at very high pressure and temperature.

There's nothing very pretty about a column still and – with the exception of Woodford reserve, where pot stills are used, and Maker's Mark, which is maintained as if it were a museum – bourbon distilleries can be ugly, industrial places. That doesn't mean there is anything second-rate about the spirit they produce, though. Such distilleries can produce a high volume of top-quality spirit, which comes off the stills at a much higher alcoholic strength than Scottish single malt spirit.

Maturing bourbon

The real magic of bourbon is in the maturation phase. Unlike in Scotland, where casks that are used have previously contained something else, in the USA only new casks made of white oak can be used. In fact, the very reason for Scotland importing so many casks from the USA is because of the abundance of redundant barrels.

Before being used for bourbon maturation, the barrel is charred, or toasted, on the inside so that the spirit can more easily interract with the oak. Once the whiskey is filled into the barrel, it is stored for maturation. Huge multi-floored warehouses, each like a small block of flats, are used for this.

Casks of maturing bourbon at a Jim Beam warehouse

Due to the Kentucky climate, the warehouses bake in high summer temperatures and are at the mercy of the harsh bitter winds that sweep from the Appalachian Mountains during the short but very sharp winter. The temperature contrast in Kentucky is extreme, from sub-zero to the high 30s celsius, and in summer temperatures can easily climb above 40 degrees. Each floor of the warehouses can vary from three to five degrees, producing "honeypot" areas, where maturation is at its best. The overall effect of such variation in temperature is to put maturation in Kentucky on fast-forward.

In Scotland, there is a dignified and stately maturation period. In Kentucky, the liquid's temperature rises and falls more dramatically than Wall Street during a mortgage crisis. The maturing spirit expands and contracts accordingly, soaking deep into the wood then shrinking back, rapidly adding flavour and colour to the spirit.

Four Roses small-batch bourbon whiskey

The flavours of bourbon

The reaction between wood and spirit produces rich fruit and vanilla notes that are instantly recognisable as bourbon. A sweet, burnt toffee character is very typical of bourbon too, combined with a medley of flavours that might include confectionery, the waxiness of saddle polish and rich cherry fruit cake.

While maturation in Scotland must take place for a minimum of three years, in Kentucky it is two. And while, as a rough guide, single malt whisky reaches its best somewhere between 12 and 18 years, bourbon is seen to achieve its premium age between six and nine years.

American rye whiskey

While by law bourbon requires a mash containing 51% corn, it is possible to make other styles of whiskey by varying the kinds of grain in the mash bill. One of the most underrated styles of whiskey, for instance, and one that is having something of a resurgence, is rye whiskey, which requires at least 51% rye in the mash bill. It's still not easy to find, but Rittenhouse Rye, from Heaven Hill, is a good starting point; Van Winkle Family Reserve Rye 13-year-old is like going three rounds with Muhammed Ali; and Sazerac Rye is a world-class whiskey, both in its youthful guise (as a 6-year-old) or in its full 18-year-old maturity.

Rittenhouse rye

Rye whiskey can be made with malted or unmalted rye, and, at its best, is big, oily and spicy – and unforgettable too.

American Whiskey

World Whisky

France's Eddu whisky

A large number of countries are producing whiskies these days, and in some cases entirely new styles are being made. Austria's Weidenauer Distillery produces an oat whisky and "Spelt" whisky, which uses a special strain of wheat. In France, Distillerie des Menhirs produces Eddu Gold and Eddu Silver, which is made with buckwheat (thereby flirting with the rules of whisky, as buckwheat isn't technically a grain). South Africa, Australia and New Zealand have all made great contributions to the world of whisky too. The inquisitive should also try Penderyn from Wales, the cask-strength version of Amrut from India and the most recent Mackmyra First Edition from Sweden.

However, there are two key nations that deserve a little more attention: Canada, once *the* dominant force in the world of whisky, and Japan...

Japanese whisky

Japan has been producing whisky for the better part of a century now, and, in recent years, has begun to reap the rewards of all the hard work and practice; since 2000, Japan has released a number of truly world-class whiskies.

Whisky production in Japan was started by Shinjiro Torii and Masataka Taketsuru. Taketsuru had worked in distilleries in Scotland, and he realised that parts of Japan provided the perfect environment for whisky production. The first distillery was Yamazaki, situated between Osaka and Kyoto. Taketsuru was responsible for Japan's second distillery too, after he left Yamazaki Distillery and identified what he thought would be the perfect location to set up his own distillery, Yoichi, on Japan's most northerly island, Hokkaido. Taketsuru's company became Nikka, and Torii's company, based at the Yamazaki Distillery, was later renamed Suntory. To this day Suntory and Nikka account for more than 80% of Japan's whisky.

Yamazaki 25-year-old malt whisky

Nikka's Yoichi Distillery on the island of Hokkaido

The Japanese whisky industry has had to overcome some hefty problems. If there has been a snobbishness from the West towards Japanese whisky, then it has suffered the same difficulty two-fold on its home turf, where Scotch whisky is revered. Added to this is the Japanese work culture, in which it is traditionally not acceptable for rival companies to interact with each other. Whereas Scottish distillers share and trade malts for their blends, in Japan traditionally no such arrangement has existed. Therefore, for Japanese distilleries to produce blends, they have either had to produce a range of whiskies themselves or bring in Scottish malt, undermining the whole concept of a unique Japanese whisky style.

In recent years, the drinks critics Michael Jackson, Jim Murray and most recently Dave Broom have championed Japanese whisky, and it has started to attract considerable interest from whisky enthusiasts. A clutch of prizes – from attaining top score in the inaugural *Whisky Magazine* Best of the Best awards in 2001 to gaining two of the top five places in the 2007 World Whiskies Awards – has confirmed the quality.

Suntory's 17-year-old Hibiki blend

Best of all from an enthusiast's point of view, the whiskies we're now tasting are not only of an extraordinarily high quality, but they seem to be developing a distinctive Japanese characteristic all of their own – a mushroomy quality. It's something of an acquired taste, but, once you buy into it, it's the clearest proof yet that Japan is capable of producing whiskies that are far more than just copy-cat impersonators.

Unsurprisingly, most of the whiskies making it over from Japan come from the two big companies, Suntory and Nikka. But it's worth seeking out whisky from Fuji-Gotemba, Karuizawa, and Ichiro Malts too. Japanese whiskies tend not to be cheap, but there is a growing band of whisky enthusiasts who find them every bit as exciting as those coming from Scotland or America. And plenty who believe that some of them are set to become major stars.

Canadian whisky

The Canadian whisky industry is like a former heavyweight boxer seeing out his twilight years, and now the glorious champion of yesteryear is struggling to survive. Nearly all of Canada's principle whisky brands are under foreign ownership these days, and the few exceptions rarely make it over the border into the USA, let alone anywhere further afield.

Yet Canadian whisky, as we know it today, has a history stretching back to the mid-19th century. Following the introduction of the column still and continuous distillation in the 1830s, distilleries sprang up across the country. And during the country's golden period in the first half of the 20th century, the mammoth companies of Hiram Walker and Seagram not only stood proud throughout North America, but in fact also dominated the entire whisky world.

Canada perfected a whisky style of its own, too, based around rye. Unlike American rye – where spicy and bold rye dominates a mash bill normally made of three grains – Canadian whisky blended corn-based spirit with a large number of different ryes produced in complex distilleries using different fermentations, spirit runs and distillations. And, under Canadian rules, a small amount of other whisky could also be added to the mix.

Canadian Mist

The resulting whiskies were smooth, rounded and easy to drink. But they were also highly complex, sophisticated and brilliantly balanced too. Good examples of this style of whisky can still be found today emanating from distilleries such as Walkerville and Canadian Mist in Ontario, and at the Gimli Distillery in Manitoba.

Key Terms

alcohol by volume (abv)
The alcoholic strength of a drink expressed as a percentage of the liquid overall. It is usually abbreviated to abv.

blended malt whisky
Confusing modern term for vatted malts – a mix of malt whiskies from different distilleries.

blended whisky
A mix of malt whiskies and grain whisky.

brewing
Process of turning a grain and water solution (mash) into beer (wash) by adding yeast and producing alcohol and carbon dioxide.

cask strength whisky
Whisky that is bottled at the strength it comes out of the cask (typically 57-63% abv) as opposed to the diluted 40-46% strength at which most whiskies are bottled.

chill-filtration
Impurities or congeners can make whisky cloudy when cold or if water is added. They can be

removed by chilling the whisky and passing it through cardboard filters. There has been a trend away from chill-filtering, though, because it is believed that some impurities can contribute to the whisky's overall taste.

coffey stills
see continuous distillation

column stills
see continuous distillation

congeners
The name of the scores of organic chemicals in the distilled spirit. Many have to be separated from the final whisky because they are poisonous or taste horrible, or both.

continuous distillation
The process of making grain spirit by forcing the wash over steam under high pressure. The process can produce large quantities of high-strength spirit. Rather than being carried out in a *pot still*, continuous distillation uses column, or Coffey, stills.

grain whisky
Whisky made from one or more grains, such as corn, wheat or unmalted barley. Grain whisky is made using continuous distillation, and a small

amount of malted barley is used in the process
to aid the fermentation.

malt

Grain that has been tricked into germinating
using moisture and warmth, then kilned to halt
the germination. Most often the grain is barley,
but rye can be malted too.

mash

The mix of grist with hot water. The sweet liquid
that results from the mash is called wort.

mash bill

Largely American term for the percentage of
each grain used to make whiskey.

mash tun

Vessel where grist and hot water are mixed.

peated malt

Whisky made with a percentage of malted barley
that has been dried over a peat fire and which
takes on its smoky and phenolic characteristics.

pot stills

Copper kettles used for distilling in batches. They
come in myriad shapes and sizes, and such
variations affect the whisky's flavour.

reflux

Process by which heavier spirit fails to reach the

lyne arm near the top of the still and runs back into the still. When reflux is high, only the lightest spirits are recondensed and a light whisky is produced. In shorter, more squat stills, reflux is low, resulting in a heavier, oilier spirit.

single malt whisky

Malt whisky is made from only malted barley, yeast and water. If it is a single malt, it is the product of just one distillery. It may contain whisky from lots of different casks, though.

spirit still

The still in which the final distillation takes place. Compare with *wash still*.

still

The vessel used for distilling spirit. There are two main types: a *pot still*, used for batch distillation; and a continuous or column still, used for *continuous distillation*.

vatting

The mixing of malts from different distilleries. Vatted malts are now referred to as blended malt whiskies.

wash still

First pot still used in distillation, producing a liquid with an abv of just over 20%.

Index

Acknowledgements

The publishers would like to thank the following for their assistance with images:

A Smith Bowman, Allied Domecq, Amrut, Bacardi, Bakery Hill Distillery, Balmenach Distillery, Barton Brands Ltd, Beam Global Spirits & Wine, Benriach Distillery, Berry Brothers & Rudd Ltd, Bertrand, The Brown-Forman Corporation, Bruichladdich Distillery, Buffalo Trace, Burn Stewart Distillers Ltd, Campari, Chivas Brothers, Clear Creek Distillery, CL WorldBrands Ltd, Copper Fox Distillery, Cooley Distillery, Compass Box, Diageo, Eddu, The Edrington Group, Fortune Brands Inc, Four Roses Distillery, Glenora Distilleries Ltd, George Dickel Distillery, Glenmorangie, Gilbeys, Heaven Hill Distilleries, Highland Park, Highland Distillers Plc, Highwood Distillery, Höhler, Hiram Walker & Sons, Inver House Distillers, Irish Distillers Group, Isle of Arran Distillers, John Dewar and Sons Ltd, Kittling Ridge, Lark Distillery, Mackmyra Distillery, Des Menhirs, Morrison Bowmore Distillers Ltd, Mitchell & Son Wine Merchants Ltd, Nant Distillery, Nikka, The Owl, Penderyn Distillery, Pernod Ricard, Piedmont Distillers Inc, Sazerac Company, Smith's, Southern Distilleries Ltd, Spencerfield Spirits, St George's Distillery, St George Spirits, Suntory, Tobermory Distillery, United Distillers & Vintners Ltd, United Spirits, Van Winkle Whiskeys, Whyte and Mackay Ltd, William Grant and Sons, Winchester Cellars.

A special thanks to Christopher Maclean and Rebecca Laurance at The Whisky Shop, London, and Malcolm Mullin at The Vintage House, London, for their assistance.

Collins Gem Whiskies